THE
LOST
FAMILY

THE
LOST
FAMILY

How DNA Testing
Is Upending Who We Are

LIBBY COPELAND

Abrams Press, New York

Library of Congress Control Number: 2019939893

ISBN: 978-1-4197-4300-9
eISBN: 978-1-68335-893-0

Printed and bound in the United States
10 9 8 7 6 5 4 3 2 1

Abrams Press® is a registered trademark of Harry N. Abrams, Inc.

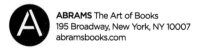

ABRAMS The Art of Books
195 Broadway, New York, NY 10007
abramsbooks.com

For Dan, always

CONTENTS

PROLOGUE

This is the story of a woman named Alice, but it is also the story of a man named Jason, and a woman named Jacqui, and a whole host of people you haven't yet met but who could be you. This is about the countless people who have seen ads for at-home DNA tests, and—eager to learn about their ancestral heritage, perhaps—decided those tests looked like fun, and got more than they expected. It could be that there's a fundamental fact about you, or about your family's past, that you don't yet know, and that will change the way you think about truth and family and who you are. The only thing standing between you and the discovery of this story is that DNA test.

If the hundreds of Americans I've spoken with are any guide, before you take such a test, you're unlikely to think there's any such disruptive fact in your family history. Even if you know there's a statistical possibility that you'll discover something surprising, you're unlikely to expect the statistics will apply to you. The problem is, it is impossible to say, in the absence of the inkling of any such surprise, whether or not you'll be better off for knowing this information. And once you know it, it's too late.

I came to think of the community of people I encountered researching this book as seekers. They are people obsessed with figuring out just what's in their genes. Not everyone who orders a direct-to-consumer DNA test qualifies as one, though a lot of folks who enter into this casually find themselves unexpectedly captivated by questions they never thought to ask, questions posed by the DNA itself. At-home DNA testing is sometimes called "recreational," to distinguish it from genetic tests that are ordered by doctors. But its implications can be far more profound than the term "recreational" implies.

The seekers generally fall into one of three categories. Some started out as avid genealogists. Before they ever ordered a DNA kit, they were already subscribers to the vast stores of genealogical records at Ancestry.com. Some had already flown out to the Mormon-run Family History Library in Salt Lake City, Utah, and spent a week there, camped out, up to their ears in old church records on microfilm. For these people, DNA testing through a company like Ancestry or 23andMe was a logical extension of this deep genealogical curiosity. Indeed, they probably tested in the early 2000s, when the tests available were much cruder and could tell you much less about your immediate forebears. Some of them, DNA junkies, have tested over the years at a variety of companies, including some that no longer exist, spitting and swabbing and scouring their results for meaning, in a grand history project of the self. These days, they canvass relatives willing to test, and return from family cruises with their kin's samples. They try to trace their family trees back hundreds of years, using DNA combined with genealogy to suss out the identity of a particular nineteenth-century ancestor or to figure out how exactly they're related to a distant cousin. If elderly relatives have difficulty generating enough saliva to spit into a vial, these seekers know the answer is a slice of lemon. A few, thwarted by the unexpected timing of events, have even resorted to swabbing the insides of their parents' cheeks after death.

The second category of seekers is marked by a more immediate and pressing puzzle—brought on, say, by lingering suspicions that the man they called "Dad" might not be their genetic father. Or they might be looking for evidence that they're not related to a family they never felt at home in. Some find their suspicions validated and others are disabused of them, and some find something else entirely, like a previously unknown half-sibling. Many seekers are adoptees looking for their biological relatives, or people conceived via sperm donor seeking the stranger who contributed half their genetic material. Finding genetic cousins in a DNA database can make it possible for a seeker to trace his way to his biological mother even if that mother never submitted her saliva for analysis. These seekers join Facebook groups dedicated to the mechanics of what's known as genetic genealogy, learning to talk of haplogroups and centimorgans and triangulation. They sketch countless family trees, trying to figure out where they might fit into them. They call their hobby an addiction. They read tips on how to phrase letters of introduction to their fathers, a topic so fraught and fascinating for what it reveals about the vagaries of human nature, and the complications of the human heart, that it deserves a dissertation or three.

The third category of people have no inkling they're about to find anything surprising, and this is often the most disruptive way in which DNA testing plays out. These people do not start out as seekers. Instead, they test because they want to find out where their forebears came from. They test to find out if that old family story of a Native American ancestor is true. They test because they want to know whether they're like that guy in the AncestryDNA commercial, who thought he was one thing and turned out to be another: "I traded in my lederhosen for a kilt!" he declares, happy in his certainty. They test because they got a kit as a gift for Christmas—and then Easter dinner becomes awkward, because the test has revealed something strange. Perhaps a stranger has appeared in the DNA database who might be an aunt or a niece or a grandparent or a half-sibling, based on the shared genetic material, and now everyone is asking questions. Perhaps a family's understanding of its heritage is muddied by a pie chart offering one's "ethnicity estimate," and the question "Why is it saying I'm half-Greek?" becomes a curiosity, which becomes a nagging doubt, which becomes a family conversation no one will ever forget. Then, these people, too, become obsessives, trying to solve the mystery, finding possible relatives through Google and Facebook searches, and then offering to buy and ship DNA kits to those people if they'll agree to test.

Well, some of them do. Some of them, presumably, back away slowly, like a hiker who spies an old grenade from a long-ago war on a beach. Those people don't tell their stories. Who could blame them? Those people don't *have* stories; they have DNA results they'd rather not have seen, and which they may prefer not to believe.

The seekers filled my inbox and my ears over the last few years with their questions. What sense could they make of a secret kept for so many decades? How much of the past did they need to go back and footnote in light of this new knowledge? How many conversations were, in retrospect, lies of omission? If old taboos—about being the child of an extramarital relationship, or the child of adoption, or being conceived via sperm donor—were a thing of the past, why was it that the shame surrounding these origins felt so palpable even now? And how *now* to treat these strangers who'd come abruptly into their lives, these strangers with whom they shared genetic material? Are these people "family," and if not, do we need a new category to describe them? Is it always better to know these genetic relatives exist? Is it always better to know the truth? Should they invite the truth over for Thanksgiving dinner?

Secrets, we are all discovering, have a propulsive power all their own, and time and complicity only make them more powerful. Once you decide to keep a secret, the secret maintains a circular logic, even when circumstances change. Many seekers say the fact of the secret is the thing that nags at them, more than the nature of the secret itself.

DNA testing has brought the past forward to the present, forcing us to grapple with decisions made long ago in different, often desperate, circumstances. It forces us to think about the people whose truths have been hushed up for decades—the teenager consigned to a home for unwed mothers, the medical student who contributed his sperm, never dreaming that sperm would become a person knocking on his inbox five decades later. Some seekers are in their eighties and nineties, just discovering siblings they've spent their lives without. Some are left with only questions, and there is no one left to ask.

"I had no clue that I was adopted," says a seeker named Linda who discovered this blunt fact at the age of fifty-one, through a DNA test. When Linda figured out the identity of her biological mother using puzzle-solving techniques well-known to seekers, she learned the woman had just passed away, and that she would never have answers to certain things. "Did she ever think of me? Did she ever look for me?"

In researching this book, I toured the first American company to offer at-home DNA testing for ancestry, in Houston, Texas, and the biggest, in Lehi, Utah. I tested my own DNA at three companies, and looked for relatives, and learned the imprecisions of ethnicity predictions while comparing the results the various companies gave me. I spent a day at the biggest genealogical library in the world, where history was a breathing thing, alive in the records. Meanwhile, police began making end runs around the limitations of government-run DNA databases, instead using genetic material gathered by consumer testing services to solve horrific cases that had been cold for decades—and igniting fierce privacy debates. And the databases of people who'd undergone at-home DNA testing went from eight million by the summer of 2017 to nearly thirty million just two years later. By the time you read this, the number could be two or three times that. The sheer girth of those numbers means that even if you don't choose to send away for a kit, it increasingly doesn't matter. Especially in the United States, where DNA testing is more popular than anywhere else, all of us are already drawn in by the decisions of other people who share our genetic material—people who, in many cases, we've never met. As bioethicist Thomas H. Murray told me, "You don't get to opt out."

This means you may become a statistic—someone who discovers a genetic fact that upends your sense of self, forces you to rethink what you know about race and religion, about your place in your family and your role in the world. Statistics are heartless that way. The more I reported for this book, the more seekers I talked to, the more I came to feel that we are embarking on a vast social experiment, the full implications of which we can't yet know.

This is the story of the seekers, who are grappling with essential questions about identity. What makes us who we are? It is the story of Alice, who did not know she would become one of them. That is, one of us.

1
ALICE IS NOT ALICE

The first thing Alice Collins Plebuch told me when she opened the door to her house was that I was tall. I am not tall.

Alice is very short. Five feet if she's lucky, she likes to say. She's in her early seventies, and her grandkids call her Grandma Nerd for her love of technology. Before we walked to her office-slash-sewing-room, she warned me not to take my shoes off, because sewing pins were scattered all over and I was likely to get them embedded in my feet. And then she led me back to her busy, treacherous, creative space, filled with bits of fabric and piles of books and genealogical research folders

Alice Collins Plebuch

arranged so delicately that if you had the bad sense to reach for an object that looked interesting, a whole stack of things might come tumbling to the floor, and Alice might or might not notice, and might or might not say *don't worry*, she'd get to it later.

She gave me a tour of her advanced sewing technology, so unlike standard sewing machines as to be an entirely different species. Atop one desk was something called THE Dream Machine 2 by Brother, a sewing, embroidery, quilting, and crafting machine so complex I'd been confounded each time Alice had tried to describe it to me over the phone. It was huge, with a ten-inch high-definition screen to choose an embroidery image and the capacity to scan and to plug in a memory stick so that you could add more. (It had been retailing for the cost of a Kia when she'd bought it a few years before, though Alice being Alice, she did not pay anywhere near that much.) Nearby were several other serious-looking contraptions, including something called a serger for executing what are known as overlock stitches, and something called an AccuQuilt, which I read is supposed to help you "reclaim your quilting joy."

Then Alice showed me the things she'd made on these machines, the beauty of which I could comprehend. By her desk was an early effort saluting a book series she devoured long before HBO made it into television: a direwolf from *Game of Thrones* with the phrase "Winter is coming." She showed me pictures of the fancy, full-body burgundy *Star Trek* costumes she'd made for herself, her eldest son, and his children, refashioning the complex rank braid each time her son told her it wasn't quite faithful to the on-screen version. On the back of a denim jacket she'd crafted an elaborate, plaintive-looking green-and-gold dragon, studded with red and silver crystals, its one sad eye outlined in shiny nuggets of gold. Just now she was working on removable pillow covers for the cushions of her couch, using a patterned turquoise fabric she'd found on deep discount. She had plans and measurements sketched out, and throwaway cloth for playing with, and as she showed me her plans, she improvised, folding the cloth this way and that, looking for the design that offered the right fit and the right give. For Alice, sewing is not a tribute to domesticity; she tends to resent domesticity, given the way she was raised, the eldest girl in a family of nine. Rather, sewing is nerdy, technical, challenging. Alice loves a challenge.

Over the years, I've thought a lot about Alice's brain—how for decades she honed certain skills involving technological efficiency and problem-solving, so that when the time came for her to answer the most important question of all, she dug in like she'd been training for it. When she was growing up, her family was not wealthy—no matter how many overtime shifts her dad worked, there were still seven kids to care for—so she put herself through college, washing dishes and tutoring kids in math and sewing her own clothes to save money. In time, she worked for a professor as a "keypuncher," putting information onto computer cards to be fed into a mainframe computer, and quickly demonstrated what a college friend describes as "a natural feel" for crunching data on those finicky machines. She got a degree in political science at the University of California, Riverside, specializing in statistical analysis, and nearly finished a master's degree in political science. In 1971, she married Bruce Plebuch, who studied philosophy in grad school and later became a lawyer. The same year, she got a job working as a clerk for the University of California, Santa Barbara.

There, Alice stumbled on a problem in need of solving. Faculty members' records were on card files, and a small staff spent an inordinate amount of time tracking and updating their salaries and academic advancements. Alice had the idea to put this information onto early computers. This set a pattern, and became the focus of her career in a series of positions in the field of information systems and data processing throughout the University of California system, where she worked with personnel files, payroll systems, and benefits enrollment, facets of working life that most of us notice only when they're clunky and broken. Early on, Alice was part of a rising cadre of women in a male-dominated environment. She asked her boss for a title and salary to match her responsibilities—"You know if you had a man doing this you'd be paying him a lot more"—and got the first of many promotions.

And she kept making suggestions for how things could be better. *It sure would be a lot nicer if we had these automated.* And, when the Internet came along: *It sure would be cool if people could access their benefits online.* All that data could be put to better use if only someone were willing to tame it, and Alice was just the woman for the job. "I'm not intimidated by lots of data," she told me. She'd always been good at math and science; as a kid, if a problem stumped her, she'd go to sleep and wake up with the answer. She can see the big picture, but she also notices the details. She told me it used to piss off one of her programmers, the way she could scroll through acres of his work and glimpse the one error that would mean a whole program wouldn't work. She could do it quickly, by picking out the "nits."

Alice has been retired for many years. She stopped working in her late fifties, the result of financial planning and good retirement benefits. But she remained an early adopter of new technologies. In 2007, before anyone had an iPhone, she flew to the Macworld Expo in San Francisco to cradle the prototype in her arms. She had the first model the very day it became available, and still does with every new iPhone. And when the thing happened that set everything into motion and changed Alice's life, this was also the result of her love of new technologies. Alice being Alice, she could not leave a puzzle unsolved, no matter how difficult it was or how troubling its implications.

She could not.

* * *

Alice had long had questions about her family. Her mother, who was also named Alice, was into genealogy and kept an old family bible from the 1840s with birth, death, and marriage notations that traced back her English roots. Alice the elder had encouraged her daughter to embrace her hobby. But sometimes daughters want to do the opposite of what their mothers tell them, which is why Alice the younger refused to indulge whatever nascent interest she had in the topic until after her mom died in 1992.

Then, she dove in. Alice found her mother's line easy to document, even in the years before lots of genealogical records were online. Her mom, Alice Nisbet Collins, was descended from Irish people on one side, and on the other from Scottish and English people, some of whom had been on this side of the Atlantic as far back as colonial America. Alice was able to follow one line of her mother's ancestors back to 1500s England. Alice's father, Jim, lived for seven years after his wife died, and as Alice updated him on everything she was finding along her mother's line, she began to feel guilty. "Because my dad had nothing—he had no history," she says.

Alice's father, Jim Collins

Jim Collins, the son of Irish immigrants, knew little of his parents, one of whom he barely remembered and the other not at all. His mother had died when he was a baby, and his father had given him and his older siblings away to a Catholic orphanage. For a long time, Jim didn't even know what year he was born; sending away for some vitals as a young man, he'd discovered he was a year younger than he'd thought. So, in her father's waning years, Alice dove into the project of Jim. She had just one image of him as a child, taken in 1914, shortly before he and his siblings were sent away. In the photo, curly-haired Jim sits on his father's lap, clad in a white baby dress, with his tiny sister and brother standing on either side. He is the only one smiling.

The Collins children, Kitty, Jim, and John,
with their father, John Josef

Some things Alice already knew. She knew that life in the orphanage had been difficult. She knew her dad was probably malnourished there, because a doctor later told Jim that this likely explained his small stature. She knew that Jim had left the orphanage as a young teenager, and that he'd lived some rowdy years he liked to describe as his "misspent youth," before he joined the Civilian Conservation Corps and then the US Army Corps of Engineers, and met and married Alice's mother. She knew Jim had loved his sister, who died as a young woman, and had not been close to his brother.

Alice and her siblings sensed that Jim's Irishness and his Catholicism were important to him; they were what was left of his identity when so much else was taken away. He cooked a wicked corned beef on St. Patrick's Day, and liked to brag to one of his granddaughters about the time someone had described him as a "thoroughbred Irishman." Jim and Alice the elder had five boys and two girls, and the family went to church on Sundays, and much of the kids' schooling was at Catholic schools, and it was "important to him that the kids be raised Catholic," says Alice's sister, Gerardine Collins Wiggins, who goes by Gerry. But for all that, Jim's knowledge of his roots was shallow. He knew his dad was from County Cork. Before traveling to Ireland in 1990, Gerry had sent away for Jim's birth certificate, hoping it would help her trace his roots. But when she arrived in Ireland with the document and the few family names her dad was able to give her, the office she'd planned to consult for historical resources had closed down, and Gerry couldn't search the old records. Looking back, there were several moments like that: moments when questions were asked but answers were not forthcoming, when the paucity of information available to Jim fastened riddles into tight knots.

Here's another. Jim was not terribly knowledgeable about his extended family, and they seem not to have embraced him, either in childhood, when he might have needed them most, or later. But in adulthood, Jim tracked down an aunt while traveling in the Southwest, and went into the hotel she was running to inquire if he could visit with her. He could hear her informing someone that she was taking her nap and that her nephew would have to come back later. He left, hurt, and never met with his aunt, and any questions he might have had about the mysteries of his childhood and the parents he barely remembered and the extended family who didn't take in those three kids bound for an orphanage—those questions went unasked.

When Alice embarked on her genealogical quest on behalf of her father, she knew it would not be nearly as easy as it had been for her mother, but that was OK. Alice likes research. She sent away for Jim's parents' death certificates and headed over to the National Archives in San Francisco, where she was living at the time, to comb through microfilm, looking for her grandparents in the 1910 census. She was surprised by how little she knew about Jim. At one point, she realized she didn't even know his mother's proper first name.

She was able to glean just the barest outline of her father's parents. When Jim was born, his family lived in a working-class neighborhood in the Bronx. His father, John Josef Collins, was listed as a "driver" and later a longshoreman. His mother, Katie, died at thirty-two, when little Jim was just nine months old. When Alice got her grandmother Katie's death certificate, she strained to read the words by *cause of death*, which were written in the kind of loopy, inscrutable script that is the bane of every genealogist's existence. She thought it read, in part, "cranial softening" or "cerebral softening"; she showed it to a pathologist friend, who thought Katie might have had encephalitis. The last word, though, was absolutely clear. It said "insanity." How, Alice wondered, does a person die of insanity?

Before Jim died in 1999, just before his eighty-sixth birthday, Alice wasn't able to tell him much more than he already knew. And in the years that followed, she continued to be stymied by the paper trail. Searching census records, she managed to figure out the name of the orphanage in Sparkill, New York, where Jim had spent his childhood. It had been called St. Agnes Home and School, though the orphanage itself no longer existed by the time Alice found it. She wrote the religious sisters located there, asking for information on her dad and his parents, and got back a letter with the barest of facts: the date Jim and his siblings were admitted, and the varying dates when he and his siblings were discharged.

Before many Americans were aware of the way the Internet was transforming genealogical research, making it possible to find ancestors in old records from all over the world, Alice was already decades deep into this hobby and had subscribed to Ancestry.com to access its vast trove of historical records. And this is why she found out as soon as the company started offering an at-home DNA test in 2012.

These days, Ancestry's database is stocked with the genetic data of more than fifteen million customers, but back then, the kind of test the company was making available to subscribers—autosomal DNA testing—was relatively new and not widely known about. Alice's scientific curiosity was piqued. If the paper trail could not help her, she thought perhaps the genetic trail could. The timing was fortuitous, as Alice thought she'd homed in on the village her father's father had emigrated from, and was considering a trip to Ireland. She got on the company's waiting list for the test.

There are two ways that home DNA testing companies collect your genetic material. One method is to have the customer swab the inside of her cheeks, and another is to have her spit in a vial. AncestryDNA uses saliva. When the company's box arrived at Alice's house, she opened up the vial and spit and spit, about ten times, until the bubbles of saliva cleared a line on the side of the tube. She closed the top, which released a preserving liquid into her saliva, put the vial in a package, and sent it back to the company. And then she waited.

* * *

Before autosomal DNA testing came on the scene, two types of genetic testing dominated the consumer market for those interested in tracing their ancestral histories. One, called Y-DNA testing, examines the Y chromosome a man inherits from his father's father's father . . . all the way up what's known as the patrilineal line. Only men can take the test, since only men inherit a Y sex chromosome along with their X (women, of course, inherit two X's). The other test looks at mitochondrial DNA (mtDNA). Whereas our twenty-three pairs of chromosomes sit inside the nucleus of the cell, the tiny mitochondrion sits outside the nucleus with its own DNA, and both men and women inherit mitochondrial DNA from their mother's mother's mother . . . all the way up the matrilineal line. Both of these kinds of testing can tell a person about one particular genetic line, but not about their maternal or paternal side as a whole, nor about the genetic heritage they inherited from their other parent.

Because mitochondrial DNA and Y-DNA are inherited more or less intact, geneticists use mutations, which happen slowly over time, to trace historical migrations and track branches on our huge family tree of humanity. Y-DNA and mtDNA testing can tell us where some of our ancient ancestors lived, but typically can't give us precise information about our immediate family trees, nor can they offer predictions about a person's overall ethnicity. Although they can reveal that you and another person share a common ancestor, it can be difficult to tell how closely related you are to that other person. However, there are instances where these kinds of DNA can be helpful for genealogical purposes. Because the Y chromosome passes from father to son relatively unchanged, for instance, it can be used in combination with a surname, which also typically passes from father to son, to trace the genealogy of a male line.

As it happens, some of Alice's family had already begun dabbling in genetic testing. Her sister, Gerry, who was also interested in the Collins family's history, had recently had her mitochondrial DNA tested, showing her deep ancestry along her matrilineal line, and Gerry and Alice had convinced one of their brothers to test his Y chromosome, in hopes of getting a sense of their father's side. But the AncestryDNA test, when it became available, was a wholly different animal, offering much more recent and comprehensive genealogical information. AncestryDNA's test looked at the twenty-two pairs of chromosomes that don't determine a person's genetic sex, known as autosomes. It considered the genetic material contributed by both parents and by their parents before them. It could reveal recent genetic ethnicity and show relatives along both the maternal and paternal sides, assuming they were in the company's database. It promised, in other words, to help Alice unravel her father's genealogical knots.

Alice waited over a month to get her results back. Finally, she got a notification they were ready. She logged on—and was utterly perplexed.

She was just 48 percent British Isles, AncestryDNA informed her. She should have been more like 100 percent, given her dad's Irish ancestry and her mom's Irish, English, and Scottish mix. Instead, the test results suggested that the other half of Alice's ethnic makeup was a mix of what the company called "European Jewish," "Persian/Turkish/

Caucasus," "Eastern European," and "uncertain." Alice emailed her sister in Minnesota, characterizing these findings as "kind of surprising." Her tone was moderate, although she did not feel moderate. She felt confused and, as the days went on, increasingly annoyed at what seemed like flaws in this new technology. "Keep in mind," said a little note at the top of her AncestryDNA estimate, "your genetic ethnicity results may be updated. As more DNA samples are gathered and more data is analyzed, we expect our ethnicity predictions to become more accurate." *More accurate?* Did that mean the estimate was mistaken? Alice tried to imagine how it could possibly be correct. Perhaps these Eastern European results indicated ancestry going back tens of thousands of years?

"I know we pick up ancient stuff," Alice wrote her sister, but if so, why was it that so many of the modern-day customers who showed up as predicted genetic relatives to her in AncestryDNA's database shared this European Jewish ancestry? "Since we know the ancestry that far back on Mom's side, does that mean it comes from Dad? Or is Mom's side not what it appears?"

This was a new puzzle, and Alice threw herself into it. She couldn't be sure whether the mistake was in the company's test or in the story she'd long believed about herself, and while she was inclined to dismiss the test as flawed, not knowing made her uneasy. The closest matches that showed up within Ancestry's database were third and fourth cousins, and she pored over their names and realized she had no idea who any of them were. She looked at her relatives' family trees, hunting for a common ancestor who might explain how her family hooked up with theirs. She got nowhere.

Ancestry had message boards, and Alice began to lurk where she saw people complaining. A lot of testers were saying their AncestryDNA results seemed to overestimate their Scandinavian heritage—results that must surely be wrong, testers wrote, because they had no paper trail, no family stories, no nothing to indicate this much Scandinavian. This was simultaneously reassuring and discomfiting to Alice: On the one hand, it seemed to indicate that the company was capable of making mistakes. On the other hand, no other customers were complaining about her precise situation: a false positive result for "European Jewish," "Persian/Turkish/Caucasus," and "Eastern European."

There was a woman on the Ancestry message boards named CeCe Moore who seemed to know more than everyone else and was answering questions. Alice googled her, found she helped run a Yahoo group called DNA-NEWBIE, and began to follow it. At first, what Alice read was above her head, but she watched and learned, observing how adoptees used their DNA results to figure out their biological families, and following links to videos showing how DNA gets passed down through the generations, refreshing what she'd learned decades earlier during a temporary stint as a biology major.

The bulk of our DNA sits in our twenty-three pairs of chromosomes, which carry the blueprints for our existence, about half inherited from each parent. Our parents, in turn, have each inherited from their parents, so Alice could expect to get about 25 percent of her genetic material from each grandparent, on average, though because of the way chromosomes are randomly inherited and because DNA is shuffled through a process called recombination, the exact percentage varies. The amount of autosomal DNA coming from each ancestor would get progressively smaller each generation back, with some ancestors failing to transmit DNA altogether as they became more distant.

The funny thing is, about 99.5 percent of our DNA is identical to that of every other human being on the planet. It's only in that remaining 0.5 percent that we get the distinctions in eye color, skin color, height, and other things we can see, as well as plenty of things we can't, and it's only a portion of that 0.5 percent that companies like Ancestry are looking at when they examine our DNA sequences. Four chemical "bases" form the building blocks of our DNA—adenine (A), cytosine (C), guanine (G), and thymine (T)—and they are arranged into pairs along the double-stranded helix structure of our DNA. (A and T pair, and C and G pair.) Testing works by reading these letters. But Ancestry wasn't looking at all of them, since whole genome sequencing involves looking at all three billion or so of a person's complementary "base pairs," and was too expensive for the direct-to-consumer market. Instead, using an approach called genotyping, Ancestry was looking at about seven hundred thousand locations along the genome—called single nucleotide polymorphisms, or SNPs (pronounced "snips")—where the bases are known to vary among human beings.

Alice's crash course in DNA taught her a lot, but it did not tell her how accurate Ancestry's test was. The Collins kids are spread out all over the country, but it so happened that within days of Alice getting her results, her brothers Jim and Brian were both in the Vancouver area. When they visited, Alice told each in turn about her findings. *Crazy*, thought Jim, dismissing the results. *Interesting but impossible*, thought Brian, dismissing the results.

But Gerry's reaction was different. "Whoa!" she'd written back to Alice's email, in huge letters, intrigued by the mystery. The sisters started discussing what to do next, and in between, they emailed about the trip to Ireland that Alice was thinking of taking. Gerry offered Alice advice from her 1990 trip, explaining the hazards of driving on the other side of the road, and telling Alice not to be surprised that most Irish people she'd encounter would be taller than their family.

About a week later, the Y-DNA results came back for their brother Jim (we'll call him Jim the younger), showing his ancient lineage along his father's father's father's line. It listed his paternal haplogroup (a genetic population sharing a common ancestor) and his subclade (a smaller group within that haplogroup). They seemed to suggest the Ancestry results might be much ado about nothing. "Today, roughly 70 percent of the men in southern England" belong to that haplogroup and subclade, the explanation from the National Geographic Society's Genographic Project read. "In parts of Spain and Ireland, that number exceeds 90 percent." This made sense to Alice—this test was tracing her father's line, after all, and everyone knew Jim Collins was Irish.

The nuances of this finding were not yet clear to the Collinses. For a host of reasons, a finding like this wasn't necessarily inconsistent with Alice's AncestryDNA results, assuming they were correct. For one thing, the Y-DNA test did not track anything from the Collinses' maternal side, so if the mystery lay with their mother, this test did nothing to address it. And the particular results Jim the younger had received actually couldn't tell him definitively whether or not he had Jewish heritage on his father's side. But in any case, Alice's brother started calling himself "Irish Jim." Back and forth the siblings went, discussing their assorted DNA results and what they all meant. Gerry pointed to her mitochondrial results, explaining that her maternal haplogroup appeared to be of Western European lineage. "I see it as a fun mystery to be solved and nothing to be worried about," Alice wrote her siblings about the whole affair. The AncestryDNA test was still in beta testing, and this made Alice especially dubious of its accuracy. She found a feedback form and sent the company what she'd later describe as a "nasty" note: Her genealogy, family stories, and life experience all indicated she was Irish, Scottish, and English, she told them, and Ancestry's science clearly wasn't up to the task yet.

In some sense, Alice was right that the ethnicity estimates of the major testing companies are flawed, in ways both practical and existential. They are estimates for a reason, based on data and techniques that are imperfect and evolving. They are flawed in another sense, because there are big questions around what precisely "ethnicity" is and whether we can ever find such a thing in a person's DNA sequence. Those were big issues, and in time, Alice would grapple with them.

But for the time being, despite her nasty note to Ancestry, she still had a queasy feeling that she might be missing something. This should go without saying, but the notion of being ancestrally Jewish was not what disquieted Alice; indeed, she found this idea intriguing, except for the implication that she didn't know important aspects of her family's history. She had traced her genealogy along both sides as exhaustively as she could. She knew—or at least thought she knew—who her parents were. None of it made sense. She couldn't drop the matter without investigating, though perhaps some people might have. She and Gerry began to speculate about what could explain the test's finding, on the off chance it was correct.

Alice knew she'd chase this question till she had an answer, but she had no way of knowing how this would become the thing she did all day, every day, taking over her retirement, teaching her an expertise in a kind of genetic detective work that a growing number of Americans are learning. She had no way of knowing how difficult it would be to solve this mystery, or how very long it would take, and how many people she'd pull into the effort—family and strangers and genetic relatives she'd never known before. She had no way of knowing that she'd be joining a community of seekers, each one of them with questions of his or her own, and with stories about the strangeness of the human condition at this, the very dawn of the consumer genomics age.

She could not know then, in the summer of 2012, that in coming days she would start to rethink that trip to Ireland. She could not know that eventually, she'd be moved to email Ancestry's feedback team with an addendum to her earlier note: an apology.

2

CRUDE BEGINNINGS

Roberta Estes got into genealogy in the late 1970s, when she was pregnant with her first child. There were a lot of unanswered questions about
Roberta's father's side of the family, just as for Alice's. Back then, genealogy was as fast as polishing rocks. You couldn't sit at your computer,
log onto Ancestry.com, and figure out your great-great-grandfather's
name in a few hours. Oh, no. You asked family members for their stories. You wrote letters, actual letters sent through the mail. You traveled
to small towns and chatted with clerks and went to libraries to scroll
through microfilm till your eyes were red. Genealogy was for the few,
the dedicated, the serious family historians. It required leisure time
and, ideally, enough money to travel. It was not for the idly curious.

"The slowness of the process was part of it," Estes told me the first
time she talked to me from her home in a rural part of Michigan outside
Ann Arbor. She had long silver hair and was wearing a T-shirt that said,
"I'm a genealogy rebel—I research past my bedtime."

It can be hard to grasp how profoundly the world of genealogy has
changed from then to now, with the advent of the Internet, social media,
and consumer DNA testing. The bar to entry is so much lower than it
was when Roberta Estes started. Since the first genetic genealogy tests
became available in 2000, DNA testing has become much cheaper and
much better at finding close relative matches. Now, for ninety-nine
dollars (and less during flash sales), the idly curious can stumble into
genealogy; within weeks of submitting saliva they'll have one of those
ethnicity estimates like the one Alice got, as well as a list of perhaps
thousands of relatives who match them in a company's database. They
can quickly find out more about those relatives through family trees,
Facebook, Google, LinkedIn, and people-search websites. If they want
to learn more about their ancestors, they need not write letters and

travel to libraries but can see billions of the world's genealogical records instantly via a subscription at Ancestry and through the free resources at the Mormon-run FamilySearch.org. And they can join groups, mainly on Facebook, for all the flavors of genetic genealogy seekers: adoptees seeking their biological families; African Americans and people of Jewish descent, who face unique roadblocks when they search; those who've discovered that the men they thought of as their fathers aren't their fathers as far as biology is concerned; those conceived by donor sperm. There are groups for those whose ancestors come from County Donegal in Ireland, and for Asians adopted into the United States, and for Americans descended from Germans who lived in Russia.

Which means that whereas in the past, you had to work years, even decades, to solve a mystery—and Roberta eventually found she had one, too—nowadays, the answer can often be found within days.

America has become a nation obsessed with genealogy. The mere existence of so many genealogical materials digitized, indexed, and searchable online, and our communal drive to find them, comes from a suite of personal and cultural motivations, as well as a complex history around the search for lineage. In his 2013 history of American genealogy, *Family Trees*, historian François Weil traces how the American impulse toward genealogy has often been in tension with itself. In the early days of the new American republic, Weil writes, the idea of establishing one's family line was associated with the British aristocracy's obsession with social rank, and viewed with suspicion by a society that saw itself as more egalitarian and forward-looking. Why would one be driven to document one's ancestors, if not to prove some connection to better birth and station?

But over the course of the nineteenth century, that shifted, enough that by 1879 the *New York Times* could declare that "we are becoming the most genealogical nation on the face of the earth." Weil writes that American genealogy transformed into a respectable middle-class endeavor as Americans began to justify and sanctify the activity within the context of family, which came to be viewed as an almost holy thing. The family "was viewed as a refuge from the outside world in an ever-changing environment," Weil writes, and genealogy became a mechanism for remembering and solidifying that unit. Besides, some Americans came to see the process of learning one's family history as a moral endeavor—a person could learn much from what her ancestors had done right or wrong. Reframed within the context of republicanism and democratic ideals, genealogical inquiry could become the means to celebrate not just the richest and most titled of forebears, but even the humbler sort. One 1850s Pennsylvanian went so far as to boast of his family's "mediocrity." The practice of keeping one's family history in a household bible had long been popular; now, middle-class New England families augmented those bibles with wall hangings of family registers and embroidered family trees.

Yet for some, genealogy was also a means of forging identity by relating to some people and excluding others. Even before the Civil War, there was "lineage consciousness" among those descended from elite colonial families, who used their descent from "high" birth to justify and enforce their higher social rank. (Some nouveau riche bought fake family pedigrees and had family crests painted on their carriages.) Americans also became intrigued by the idea that the country boasted different "stocks"—there were the Puritans, the Southern Americans, the New Yorkers. And in the wake of the war, Weil writes, genealogy became a vehicle for many whites to bind "ancestor worship, nationalism, and racism" into one big ugly package. By the early 1900s, these notions were well entrenched, so that presidents general of the Daughters of the American Revolution could talk of the importance of the country's Anglo-Saxon roots, of the "purity of our Caucasian blood," and the importance of "the white man's standard of living."

Of course, there were other forces at play in this transformation, including some key scientific breakthroughs that would form the basis for the field of genetics—as well as a theory that, in attempting to explain the vast social, racial, and economic divisions that ran through society, served to reinforce them.

* * *

Gregor Mendel died in obscurity. The priest and naturalist whose work helped us understand genetic inheritance was only recognized well after he died. Born in Moravia (now part of the Czech Republic) in 1822, Mendel made his breakthrough in the 1850s and '60s, breeding garden peas in the monastery garden at Brno. In a series of innovative experiments, he examined things like height, seed shape, and pod color, trying to track the traits of plant hybrids through the generations. Years of study led him to develop a theory of dominant and recessive traits that explained how things like wrinkled seeds can seem to disappear for a generation or more, and then reemerge intact.

Mendel's intention was not to establish laws of heredity, nor to anticipate the concept of the gene, but when his 1866 paper on those pea experiments was rediscovered in 1900, he came to be seen as a prophet because of all that had happened in the sciences in the decades in between. In 1859, Charles Darwin had published *On the Origin of Species*, describing his astonishing theory of evolution by natural selection. *Origin of Species* explained how selection drives change by way of inherited variations within populations, helping organisms with traits better suited to their environments to survive and pass their adaptations on to their young. The book was seen as intriguing, radical, implausible; copies flew off the shelves. Although Darwin's ideas weren't immediately accepted whole cloth by his peers, they sparked larger conversations about the concept of evolution. There were also lively debates about the mechanism through which heredity might work, though none of those theories—including Darwin's, involving something called "gemmules"—would prove right.

Origin of Species had a big influence on Darwin's cousin Francis Galton, an explorer, statistician, and all-around polymath who was fascinated by the topic of heredity. Galton became deeply interested in studying the relative contributions of what he would later coin "nature and nurture," helping frame a debate that has fascinated generations of psychologists (and would, over a century later, entrance a six-year-old named Anne Wojcicki, who grew up to co-found a company called 23andMe). Galton was a data guy—known to methodically count yawns during lectures—and he set about quantifying what made human beings what they were, pioneering methods that would become influential in human genetics research. He had the innovative idea of using twin studies to examine the influences of nature versus nurture. He used pedigree charts to try to study how traits ran in families. He set up an "Anthropometric Laboratory" at a major exhibition, charged visitors three pence for admission, and collected information on all sorts of things about them, like "keenness of sight and of hearing," "force of blow," and "breathing power." He developed a system for classifying fingerprints and studied photographs of criminals, hoping their facial features would illuminate their characters.

Ten years after Darwin published *Origin of Species*, Galton published *Hereditary Genius*, attempting to demonstrate through the genealogical study of eminent men that character, intellect, and ability ran in families. (He dismissed the idea that these judges, statesmen, and commanders might have benefited in a big way from social and economic advantages.) And he offered his theory of how to make humanity better: Just as it was possible to breed horses and dogs for speed, people could be bred through a series of "judicious marriages." Later, he coined the word "eugenics" to describe what he called the "science of improving stock . . . to give the more suitable races or strains of blood a better chance of prevailing speedily over the less suitable."

All of this—as well as a better understanding of cells and new thinking about the nature of heredity—created a new context for the rediscovery of Mendel. His paper on pea experiments, seen as proof of the existence of paired hereditary particles, set the stage for more research establishing the classical concept of the gene as a material unit in the chromosome. Of course, Mendelism and the early era of genetics tell only the beginnings of what we understand about heredity today; the rise of molecular genetics and a century of research were to follow. Mendel's pea plants could not account for the full complexity of inheritance patterns—for the role of environment, or the way that multiple genes play parts in influencing most traits and diseases. Scholars like Gregory Radick, a historian of science, have pointed out that the emphasis on the work of Gregor Mendel in contemporary science curriculums may encourage belief in "the primacy of the gene" at the expense of a more accurate modern understanding of the complex roles that genes play in shaping traits. Such simplistic thinking can contribute to genetic determinism, a sense that genes are fate.

We know this danger because we've seen it before. The rediscovery of Mendel's laws coincided with political and social movements advocating this kind of genetic fatalism—movements influenced by Galton, whose work lent itself to a hereditarian worldview. In the early 1900s, the theory of eugenics, coupled with the principles of Mendelian genetics and the reform-minded zealotry of the Progressive Era, became a frightful cultural force, the basis for a supposed biological justification for racism and discrimination that spread through England, America, and much of Europe, and still echoes in far-right corners of the Internet today. At the First International Eugenics Congress in 1912, lecturers showed pedigrees appearing to document families doomed to imbecility and lunacy. Germany, whose scientists referred to eugenics as "race hygiene," was ominously well represented at the exhibition. In the United States, Indiana enacted the first compulsory sterilization law in 1907, with more than thirty states passing similar laws over the coming decades.

To modern eyes, eugenic efforts can appear by turns horrifying and comically unscientific. The concept of "eugenic love," which promised to scientifically modernize marriage by encouraging what one editorial called "love in the right direction," became popular among women's rights reformers. At state fairs across the United States, Better Babies Contests emerged, where babies could be examined and tested, with points taken away for defects like scaly skin, delayed teething, and slow reactions. Meanwhile, psychologists administered IQ tests to US Army recruits and concluded that "race" was linked to intelligence; conveniently, the racial hierarchy they discovered perfectly conformed to preexisting prejudices. Eugenics influenced the Immigration Act of 1924, which restricted the numbers of people coming from "socially inadequate" areas like eastern and southern Europe.

In other words, eugenics was no mere fringe movement; its assumptions shaped what both scientists and reformers thought they were seeing. There was panic about the degradation of the human race by the unfit. "It is difficult to find many early-twentieth-century American biologists who were not advocates of eugenics in some form," observes historian Mark Largent. In 1927, in *Buck v. Bell*, the US Supreme Court upheld a state's right to forcibly sterilize a woman after experts testified that not only was she "feebleminded," but her mother and infant daughter were, too. "Three generations of imbeciles are enough," Justice Oliver Wendell Holmes Jr. famously wrote.

In Germany, where the Nazis explicitly borrowed from American eugenicists, the movement reached its terrible apogee, with forced sterilizations, marriage restrictions, euthanasia of the mentally diseased and disabled, and, finally, genocide. The Supreme Court's decision in *Buck v. Bell* was actually invoked at the Nuremberg trials in defense of Nazi sterilization experiments. German atrocities provoked revulsion in America, but even so, Largent argues, it took decades for the movement to fade, and legislation to limit immigration—justified by eugenics—persisted. Involuntary sterilizations also continued, about sixty thousand between 1907 and the 1970s, when the laws authorizing them began to be repealed by states. Yet as these ended, historian Alexandra Minna Stern told me, a different set of circumstances enabled another wave of unwanted sterilizations affecting women of color and poor women. Reporting and a state investigation revealed that California was unlawfully sterilizing female prisoners as late as 2010.

Historian Nathaniel Comfort argues that eugenics is not an aberration of history. He hears its echoes in the promise of medical genetics, in thinking that frames "heredity as the essence of human nature," in our impulse to perfection via the genes. The legacy of eugenics is just part of the much larger story of how genetics has helped us understand where we came from and how we're made, but this legacy matters when we look at all the questions that ancestry testing and genetic genealogy bring to the fore: How do we define what we are, and what role should genetics get to play in that definition? How much of what we read in our genes and in our lineages is a product of the very specific culture we swim in? Today, some scientists debate whether "race" as a proxy for genetic variation and difference is useful, if imperfect—or so flawed that it's not a worthwhile scientific concept. (We'll return to this later.)

We look for ourselves in our family histories and in our genes, but such things alone do not make identity. We human beings are the meaning-makers, each of us a product of a particular time and place, with ideas about what we value and, indeed, what we hope to find when we look.

* * *

I went to look, too. Partly, I was interested in how genealogy informs how we think about ourselves—and what it is that made people like Roberta Estes and Alice Collins Plebuch seekers so many years back, before there was such a thing as DNA testing or online census records, when genealogy required patience and spare time. I went where seekers go. I went to Salt Lake City, to the Church of Jesus Christ of Latter-day Saints, also known as the Mormon Church.

"The greatest responsibility in this world that God has laid upon us is to seek after our own dead," Joseph Smith, the founder of the Latter-day Saint movement, said shortly before his own death in 1844. The charismatic figure experienced his first religious vision at the age of fourteen on his family's farm in western New York, and founded a movement that eventually moved to Utah. Smith was aware that the young religion had a problem when it came to the souls of loved ones who'd died before it was established: How could these people be saved if they'd never had the chance to accept the restored gospel?

As recounted in the LDS Church's history of its genealogical efforts, *Hearts Turned to the Fathers*, Smith had a vision in which the Lord told him that death was no barrier to salvation for the soul of a person who would have received the Gospel when alive. After that, he preached the doctrine of baptism for the dead. In fact, the Church came to believe a series of practices were necessary for salvation, and that it was the sacred duty of church members to perform them on behalf of their ancestors. These included not only baptism by proxy but also the enact-ment of temple "sealing" ceremonies that create eternal bonds within families. It matters deeply to members of the Church of Jesus Christ of Latter-day Saints who their ancestors are because this knowledge allows them to carry out a grave responsibility: to save souls and ensure the sanctity of the family unit in the afterlife. At times, the sense of mission among members have of the LDS Church has gotten out of hand, as when members have submitted the names of Jews, including Holocaust victims, for posthumous baptism.

That genealogy is not a mere hobby for the Church helps explain the enormous resources it has poured into the gathering and dissemina-tion of old family records. In 1894, the Church created the Genealogical Society of Utah, which is now known as FamilySearch International, the Church's nonprofit genealogical arm. By the late 1930s, the Church had begun microfilming archival records all over the world, and in 1965, it completed a storage facility called the Granite Mountain Records Vault, meant to protect all those records from decay and disaster under more than seven hundred feet of stone. Weil writes that it was also around this time, during the latter half of the twentieth century, that genealogy in general began to shed some of its racism and nativism, helped along by the civil rights movement and by Alex Haley's novel and miniseries *Roots*. The pursuit of family history became a means for a wider, multicultural swath of Americans to understand who they were and how they got here. In more recent decades, the personal computer and the Internet made family history research easier and far more acces-sible to the average American, as well as an increasingly big business.

FamilySearch, though, is not a business. It is a massive project free to everyone, dedicated to the idea that we're all better off if we know our ancestors. Across the world, it maintains more than five thousand fam-ily history centers, what journalist Christine Kenneally calls a "sacred municipal library system." It also runs the website FamilySearch.org, where you can search more than five billion records.

But there are some things you can find only if you come to the Church's Family History Library, which is the biggest genealogical research facility in the world. The place is astonishing. Inside, different floors house regions of the world. There are records from more than 120 countries, in more than 170 languages, including the largest collection of Chinese genealogies outside mainland China. There are hand-drawn pedigree charts from the Polynesian state of Tonga, and old and histori-cal books with names like *Rural Cemeteries, New Canaan, Connecticut*. There is special seating for the deaf and people with disabilities, and brochures in fifteen languages, and rows and rows of microfilm readers. Over the course of a week, the library hosts as many as four hundred volunteers, in addition to paid staff. Some are retired and on multi-year missions, having paid their own way from their homes to live nearby and help out here. One library researcher is an expert in old-world syna-gogue records, while a volunteer expert is skilled in deciphering Gothic German script, a disappearing wisdom valuable to visitors who show up with their family secrets encoded in old German church documents. And what's here is just a fraction of the LDS Church's effort to preserve the past. FamilySearch has teams in more than forty-five countries working on digitizing records, though making them searchable is a whole other ball of wax, requiring the help of hundreds of thousands of people to read through old documents, decipher them, and type in salient details like name, date of emigration, date of death.

Every morning, seekers line up outside before the Family History Library opens at 8 A.M. They come with dolly carts and heavy bags filled with family photographs and document files. Paul Nauta, the library's PR director, calls these the homesteaders, for the way they set up camp at particular computers and squat all day. They may be hobbyists or pros; they may travel as groups of genealogical societies, the better to swap stories and resources. They may come from far away—Canada, France, England, New Zealand, all over the United States—and park at the library every day for a full week. Sometimes, people planning to do just a little research stay far longer than they meant to, as if they fall into some kind of wormhole that alters time. This place can do that to you.

The other major force on America's genealogy scene is Ancestry, located just half an hour away from the Family History Library. Ancestry makes money offering subscription access to old records, many of which not even FamilySearch has. Even though these two organizations are motivated by different things—the one pursues profit, whereas the other sees its work as a godly endeavor—they often team up, though they have no financial relationship. More and more, FamilySearch has been working with commercial genealogical partners like Ancestry, as well as Israeli-based MyHeritage and UK-based Findmypast, to share resources and data "because even collectively, we can't do it all," Nauta told me.

This revolution of information about the dead is breathtaking and makes you think about all the ways we're giving up our privacy not only in this life but in the hereafter. Nauta told me he was bullish on the amount of information we citizens of the Internet share online—he figured it would make the lives of future genealogists so much easier.

"Can you imagine having just a week's worth of Pinterest posts from your grandma?" he asked, and I shuddered inwardly, thinking of all the things I'd posted on social media over the years. But the concept of privacy has radically changed for all of us, even over the last decade, from what we share online to how faithfully our movements and shopping habits are tracked by corporations, to, yes, the very secrets of the genetic material curled into chromosomes in the center of our cells. Why should the privacy of the dead be any different?

Members of the LDS Church I spoke with at the library had a profound sense of connectedness to the past. The nineteenth-century notion of knowing one's ancestors as a moral endeavor was alive in them. Nauta told me that DNA testing had lowered the age of people interested in family research, and brought in adherents through a different door—they tested first and then became interested in genealogy. That was the case for me. I'd tested at 23andMe in 2017, after my dad gave me a kit for the holidays, as so many Americans do these days. The test made the past accessible in a way it never had seemed before, even if I found the results only mildly interesting: I wondered how I was related to a marvelous fantasy and sci-fi writer who showed up among my relative matches, and pondered the lost ancestor who must explain my 1 percent Korean. I was at once disappointed and relieved not to find any big surprises in my results. By then, I'd begun to talk to other testers, and to understand that surprises can cut both ways, and that boring results can be a blessing.

When I met with Nauta, I brought a piece of paper with the names of my eight great-grandparents. He practically rubbed his hands in excitement over my paternal side—"We have the most comprehensive records for Sweden of anywhere outside the archives of the country itself!"—but I told him what I really wanted was help finding the paper trail of my mom's mom's mom, Sare Kahn. My mom is descended from turn-of-the-nineteenth-century Eastern European Jewish immigrants, whose documents are far more difficult to access than those of many of my dad's great-grandparents, who include Brits and Swedes who benefited from better record-keeping. Plus, my mom's family alternated between Hebrew, Yiddish, and English versions of their names, and when they came over, many said they were from Russia, a vast area that on a modern map might be a number of different countries. The problem of using a paper trail to find your ancestors is that it favors the side less affected by war, famine, dislocation, and persecution—in other words, the people who were already privileged by history are privileged again by the records, as African American genealogists know well. In some ways, then, DNA testing is democratizing for modern-day genealogists. It can give testers access to new cousins whose family stories and family trees can provide an end run around those historical gaps.

But if there were records to be found, I knew, this was the place to find them. A library manager showed me around the Family-Search resources, and I planned to do an hour's worth of work before driving back to my Airbnb. Yet a funny thing happened after I was seated at a computer, with access to all those records right at my fingertips. Time seemed to collapse. Three volunteers guided me through all the different resources, teaching me how to follow Sare through the records like a hound following the scent of a hare. They saw patterns

I didn't, skipping over records I would have wasted time on to zero in on ones that at first seemed unpromising. They were patient. One studied the hundred-year-old scrawl of a ship's manifest until she could decipher a "J" that looked like no "J" I'd ever seen. Sometimes my great-grandmother was Sare and sometimes Sara and sometimes Sarah, and sometimes Kahn and sometimes Cohen, until she married and became a Jaffe. I found images of her marriage and death certificates, giving us for the first time her parents' names and her husband's parents' names, and I found a record of the birth of her first child, a sibling of my grandmother's who must have died young because my mom never knew that child existed. I thought I had a sense of what it must have been like for Alice and Roberta in the early days of their genealogy hobbies. I could understand what hooked them—the tantalizing hope that somewhere in the next file was a record that could explain who their ancestors were, what their lives were like before and after arriving in America.

I was so engrossed I didn't care when dinnertime passed, and at 8 P.M. I looked up in a panic and realized the garage I'd parked in might have closed. I gathered my things and rushed to the lobby, where a friendly volunteer directed me to the exit. It had been five and a half hours.

A wormhole indeed. This is how seekers are made.

* * *

The questions started for Roberta Estes when she was eighteen months old, though she wouldn't know it till later. That's when her father was in a car accident, and the hospital called his wife, and then the hospital called his other wife. Both wives showed up at the hospital with babies, which is how they found out about each other. Estes's mother dumped her father.

Roberta did not unravel the full story of her father's other family till she was an adult, and the questions sent her digging. It was well before the DNA age, so all she had was word and paper: her reticent mother's accounts, interviews with family, troves of family letters. She wanted to find her half-brother, the little boy she'd crossed paths with in the hospital when both of them were too young to know what was happening. His name, she discovered, was David Estes. She tried city directories, obits, marriage records, and divorce records and eventually consulted a private investigator, who found a police record for her half-brother and, through that, located the man's address. Roberta wrote him a letter.

David Estes turned out to be a long-haired, tattooed long-haul truck driver, and Roberta was struck both by his rough-around-the-edges demeanor and by his loyalty to those he loved and those who needed his love. The first time they spoke by phone, his deep voice was everything to her, and at first, she could barely speak. They became close. Dave was the kind of guy who rescued dogs that had been abandoned at truck stops, and once saved a Rottweiler puppy by slugging the guy who was abusing it (hence, that police record). He lived in Ohio, about three hours from Roberta, but he would take the worst runs—Idaho potato runs, which required traveling through the mountains—just because they took him near her house. Roberta loved it when David called her "Sis." She was his only biological family aside from his children. He did not talk about feelings easily, which made it all the more meaningful when he told her he loved her.

Over the years, Estes gained a lot from genealogy. She learned she was descended from a white man who owned a slave named Harriett and fathered children with her, and she found descendants of Harriett, distant cousins with whom she grew close. She learned that, through her paternal side, she had ancestors from sub-Saharan Africa. And she learned a lot about her father that was difficult to reconcile: how he drank, how he squandered his intelligence, and how he deceived two families. As a child, Roberta had idolized him, much to her mother's annoyance, and as an adult, she had to honestly look at the whole man. But genealogical research also allowed her to peel back the context for his life: The son of an impoverished bootlegger who was fed alcohol when he was hungry, he went into the hospital at least twice in an attempt to stop drinking. She wonders whether his pain may have driven him to take his own life in a second car accident, when Roberta was seven.

"Both David and I unquestionably knew that man loved us," she told me. "I have really good memories with my father, and so did David. There is just no black and white; there's just a really big range of gray."

Genealogy was an incredible gift to Roberta Estes. It solved mysteries; it gave her empathy and understanding for life's grays. And most of all, genealogy gave Roberta her brother David.

3

SOMEBODY OUGHT TO START A BUSINESS

In the summer of 2012, not long after she got her strange DNA results, Alice and her sister Gerry came up with a plan. If the Ancestry results were wrong, the surest way to disprove them was to test through another company. Alice researched and learned about 23andMe, a company that offers DNA testing directly to consumers, with a focus on genetic diseases, traits, and ancestry. She sent away for a spit kit.

Gerry decided to test at 23andMe, too. She was curious about her own results, and besides, if she matched Alice, that would eliminate some possibilities that Alice needed to consider—that Alice had been adopted or that Jim Collins wasn't, biologically speaking, Alice's father. These were contenders for an explanation, even if Gerry told Alice she was insane when Alice suggested their mother could have had an affair.

Their mom had had "eight live births in ten years," Gerry pointed out—there are seven living Collins kids, plus little Nancy Jane, who would have made three girls in the family if she'd survived past the age of two months. Gerry tried to imagine her mother carrying on something torrid in between being pregnant, giving birth, and caring for the endless stream of infants. "There was no time."

In truth, Alice thought the scenario unlikely, too. For one thing, she thought, she looked like her siblings. All the Collins kids seemed to have their father's small, hooded eye shape, and plus, all were short, like both of their parents. (They are a family for whom each inch counts; once, when Gerry declared herself five foot one and a half, Alice suggested she was using her hair to gain a height advantage.) For another, Alice could not imagine her difficult and rather unglamorous mother ensconced in an affair. When she mentioned the theory to her husband, a taciturn man with a dry sense of humor, Bruce just laughed and laughed. But the possibility could not be ignored. Jim Collins served in the army for more than twenty years before retiring at the rank of chief warrant officer 2, in 1957. He served at West Point and in Guam, Virginia, Germany, and Texas. For the most part his family traveled with him, but still, there were times when it was possible, at least in theory, that her mother could have been with somebody else.

Besides, Alice knew, the simplest explanation is the most likely one. She was noticing, when she shared her strange results with people she knew, that they tended to ask if she was certain her father was biologically related to her. It was "the most probable likelihood," her best friend, Kathy Long, observed. Alice's methodical mind could not abandon considering such an obvious solution, if only to prove it wrong. Same for the adoption theory, which seemed even more far-fetched but was still within what Alice considered "statistical possibility."

If the Ancestry results were correct, there were other theories, too—enough theories, in coming days and weeks, to fill a season's worth of *The Young and the Restless*. Some, more far-fetched, were nonetheless more palatable than the idea that Alice the elder had hidden that Alice the younger had come into the world by way of an affair or an adoption. For example, Gerry wondered whether their father's mother, Katie Collins, had had an affair. She studied pictures of her paternal grandfather, John Josef, whose picture hung on her wall, and thought the man looked nothing like their family. In truth, she'd long thought that. But if grandmother Katie Collins *had* had an affair, the mysterious lover should account for only about a quarter of Alice's ethnic makeup. AncestryDNA was telling Alice that half of her ancestral heritage was unaccounted for. Perhaps, Gerry speculated, spinning out the theory, Katie Kennedy Collins was not really Irish, but an Eastern European woman with ancestry from the Caucasus who had an affair with a Jewish immigrant. Alice tried to make the theory work. *Maybe she was adopted by an Irish couple,* she wrote in an email. *Maybe she changed her name when she came to the United States.*

It was a weird, liminal time. The sisters stewed in the questions, in the unknowns, in the riddles posed by a test that Alice would later say she'd entered into "blithely" and was thoroughly unprepared to consider.

What if she wasn't related at all to her siblings? What if everyone but her had Jim as a father—and her father was some stranger? It is a strange thing to find a mystery sitting inside you. It is a strange thing to look in the mirror at the face you've grown old with and find you don't quite recognize it.

Meanwhile, Gerry attended an Irish wedding for a friend's daughter. There were Irish songs, and there was an Irish blessing, and the priest's homily went on and on about Irish pride. All her life, Gerry had connected with her Irishness, seen it as part of her very being. As she sat through the happy occasion, listening to the singing, Gerry felt a sadness overtake her.

* * *

Genealogy had been Bennett Greenspan's thing since he was a kid. Actually, genealogy and a certain entrepreneurial streak were two of his defining characteristics when he was growing up in Nebraska, though it was only decades later that they'd come together. Greenspan is in his sixties now, but his fascination with ancestors began when he was twelve and his grandmother died. Relatives gathered for the Jewish mourning period known as shiva, and he went around asking family history questions and writing it all down. When I visited Greenspan in Houston during the summer of 2018, he showed me a family chart he'd written in 1967 or so, when he was a teenager, attempting to trace four generations back. There were corrections of misspelled names, dates filled in with different pens over time, and question marks for mysteries he could not solve. I was struck by the care he'd taken with this document, and the fact that he'd saved it. It's a rare fifteen-year-old who takes an interest in his Hungarian great-grandparents.

Around the same age as he caught the genealogy bug, young Bennett figured out how to game the junior high chewing-gum business. His father knew a wholesaler, which meant he could buy a case of twenty packs of gum for less than four cents a pack. He sold them for a dime a pack, or two for fifteen cents, making a profit even while cutting a deal. He showed up each morning at school with his pockets and socks stuffed full of gum.

Greenspan grew up and became an entrepreneur who pursued genealogy in his spare time, putting it on hold whenever life got in the way or he hit a mystery the paper trail couldn't solve. By the late 1990s, he was in semi-retirement, having recently sold his photographic supply company, and when he kindly offered to rearrange his wife's pantry—the various tomato products, he observed, were nowhere near one another on the shelves—she informed him that he needed to get a hobby.

So it was back to genealogy, and just a matter of time before Greenspan hit another question he could not answer. Genealogists refer to such frustrating experiences as "brick walls," and there are countless books and blog posts dedicated to figuring out ways around them. Greenspan's brick wall materialized when he came across a potential relative with an unusual last name he recognized from family records; the guy lived in Argentina and claimed to hail from the same Ukrainian village as Greenspan's ancestors. Greenspan was in touch with this man, but none of the historical documents he painstakingly located—an exit passport, an old marriage contract—could confirm whether they were actually related. The paper trail had failed him. And that's when he started thinking about whether science could help.

By then, a few academic papers had been published on genetics and genealogy that intrigued Greenspan. One used DNA testing along the Y chromosome to study Jewish men who believed themselves to be descendants of the first Israelite high priest through their father's line. It found that many of these men, who often carry variations on the last name Cohen, shared certain genetic markers. Greenspan realized he might be able to use the same technology to figure out if the man in Argentina was really a long-lost cousin. But he couldn't do it himself; he wasn't a scientist. He found the name of a University of Arizona geneticist, Michael Hammer, atop the Cohen paper, located Hammer's home phone number online, and called him up on a Sunday. Greenspan was excited, but Hammer seemed less than enthusiastic about taking this unexpected call on his day off.

"Look, all I want to do is buy a couple DNA tests from you," Greenspan told Hammer. Hammer told him the university didn't offer such tests to the public, so Greenspan asked him who did. The answer was nobody did.

Greenspan paused, unsure what to say, and in that pause, he told me later, something magical happened. Hammer offhandedly remarked that somebody ought to start a business, because he got so many phone calls from crazy genealogists asking him the same question.

Greenspan recognized the opportunity instantly. A split second later, he asked if Hammer might be interested in perusing a business plan that he was planning to write up—a plan to have Hammer's lab supply and analyze the tests and have Greenspan market them. Hammer said he was willing to take a look. Greenspan wrote the plan and then stalked Hammer, calling his office again and again till he could get the scientist on the line and find out if he'd read it yet.

This was the start of what would eventually be FamilyTreeDNA, the oldest American genetic genealogy testing company. "Who knew it was going to become this huge success?" Hammer remarked when we spoke. He confessed that he initially thought the company such a long shot, and was himself such a crummy businessman, that when he agreed to be involved, "I made a very poor deal with Bennett."

Greenspan and his co-founders launched in the spring of 2000, rolling out tests analyzing Y-DNA and then mitochondrial DNA. It was a heady period. At the same time as Greenspan was opening his company, multibillionaire Mormon philanthropist James LeVoy Sorenson was launching the Sorenson Molecular Genealogy Foundation, which had as its stated goal the aim of connecting the world by gathering genetic samples from around the globe and showing how similar we are to one another. Just a few months after FamilyTreeDNA launched, President Bill Clinton announced that a rough draft of the human genome had been sequenced as part of the Human Genome Project, a massive, multinational effort to decipher the more than three billion base pairs that make up the blueprint for a human being. "Today, we are learning the language in which God created life," Clinton said.

So much has changed since 2000. When FamilyTreeDNA launched, Greenspan's tests cost several hundred dollars each, and they could not tell you anywhere near as much as they can now. The company kept improving its tests as technology got better and cheaper. In 2005, the National Geographic Society launched a nonprofit anthropological effort called the Genographic Project, soliciting genetic samples to trace the ancient migration routes of people's ancestors. Two years later, 23andMe—named for the number of chromosome sets in the human genome—launched its "Personal Genome Service," offering its first kits for the whopping price of $999 and hosting so-called spit parties to gin up interest. But the real breakthrough for genealogists came in 2009, when 23andMe announced that it was able to use autosomal DNA testing to help people find their genetic cousins. These days, FamilyTreeDNA's most popular offering is autosomal testing, but it is the only major company that still offers separate mitochondrial and Y-DNA testing, a niche market.

Greenspan's first customers were serious genealogists like himself. He told me that in the beginning, that's all he imagined his market to be. He saw this venture merely as another advance for hobbyists, akin to the conversion of paper records to digital ones. At genealogy conferences, he struggled to explain the relevance of his new company to the people walking by. Sometimes, he told me, he'd literally strong-arm potential customers. He'd shake their hands and walk backward, pulling them toward his booth, talking all the way.

* * *

When Alice and Gerry sent away for kits from 23andMe, that was only part of their approach to solving the mystery of Alice's strange AncestryDNA ethnicity estimate. Thinking ahead to getting her second set of genetic results, Alice knew that 23andMe might conflict with Ancestry's findings—or it might replicate them. If the second test turned up the same unexpected ethnic mix as the first, she hoped that the company would give the same results to Gerry, because that would eliminate the theories that Alice was the product of an adoption or an affair. But if it did, the sisters would have a new problem: They would not know whether their maternal or paternal line was contributing the unexpected ancestry, and would have no clue which side of the family to focus their efforts on. Dad seemed the likelier candidate given several factors, including how little the sisters knew about his family, and the fact that Alice thought she was seeing DNA cousins from her mother's old New England side in the AncestryDNA database. But they could not know for sure. Gerry's mitochondrial DNA results couldn't rule out their mom. The solution, they realized, was to test both sides of the family.

The sisters decided to approach two first cousins—their mother's nephew and their father's nephew—and see if the men would be willing to undergo DNA tests themselves. As it happens, they had only one first cousin on their father's side. His name was Pete Nolan, and he was the son of their dad's sister, Kitty. And the crazy thing is—for as much as Alice came to love him, and as important a role as Pete would play as the sisters unraveled their mystery—he very nearly hadn't existed for them.

Pete lives outside Charlotte, North Carolina, and we spoke by phone one day. He put me on speakerphone, and I struggled to hear him through the phone's echo, and he struggled to hear me because he's hard of hearing, so his wife listened in, translating when one of us couldn't understand the other. Pete is in his upper eighties, and he shares a birthday with Alice, except that he was born fifteen years before her. The Collins kids grew up hearing about their cousin, but never having known him, because by the time they were born, he had disappeared.

Pete grew up in the Kingsbridge section of the Bronx. He was reluctant to delve into his childhood, but the paucity of what he took from that period in his life was moving to me. He was five when his mother got sick with what he'd later learn was bone cancer, seven when she died, and "the family came apart when I was nine," he said. "I was sent into foster homes." He grew up with little knowledge of his mother's side of the family, and few memories to draw upon when he grew older and wanted to find out more about it. But he remembered his mother's maiden name was Collins, and he remembered the address in the Bronx where he'd spent his early years, because his mother had made him memorize it. He told me he also remembered two uncles—his mother's brothers—who used to visit him.

Pete grew older, went into the Air Force, got married, and moved on with his life. But after he retired from his work at IBM, he tried to learn more about his mother and her family. He found his old childhood address on a map of the Bronx neighborhood, and traced a route he'd taken as a child—down the hill, cross the street, turn right—in order to figure out the name of the church where he'd attended Catholic school. He contacted the church and got information on his baptismal record from 1933, which listed his godfather and uncle, James "Jim" Collins. Searching for Jim, he wrote to the Department of Veterans Affairs and the Social Security Administration and even cold-called some men by that name. He got nowhere until, through a chance conversation with a high school classmate he reconnected with at his fiftieth reunion, he managed to figure out the name of the former orphanage in Rockland County, New York, where his uncles had spent their childhood, and in 2001, he wrote a letter to the religious sisters running the site, asking for more information.

As it turned out, this was a fateful letter.

Unbeknownst to Pete, Alice had also written to the former orphanage months earlier as part of her genealogical research. She'd been trying to find out more about her dad's childhood, and had gotten back a letter from the archivist with the barest of biographical details. But whoever was managing the mail realized that two different people were writing to inquire about the same family. Pete got back a reply letting him know about Alice. And that was how Pete found his mother's family, and came to know seven first cousins who, for most of his life, he hadn't known existed.

Alice was astonished to get a letter from Pete. He had figured in Collins family stories like a ghost. Growing up, she was reminded of her first cousin every year on her birthday, when her father would tell her his nephew, Pete, had been born fifteen years earlier on the same day. Alice's parents had even met because of Pete—Alice the elder was friends with Pete's mom, Kitty, and babysat for Pete. Jim was fond of his sister and he spoke with affection of her in later years. She was tall and willowy, Alice told me, showing me a picture of a beautiful brunette in a smart outfit. After Kitty died young, Jim fell out of touch with his sister's widower and with his nephew. Alice felt that the mystery of Pete's fate weighed on her father. She remembered him speaking about his nephew in 1999, on the day he went in for a hospital procedure that he did not survive.

"I wonder what happened to Pete," Jim had said.

Now Alice knew the answer, and both she and Pete were grateful for this connection between long-lost families. Over time, they got to know each other, and shared photographs and stories from their lives. They met in person, and Alice was surprised by how tall Pete is. (Granted, many people seem tall to Alice.) "She is the best cousin I ever had," Pete told me.

Were it not for those crossed letters, Alice and Gerry would not have had a paternal first cousin to ask for DNA, and without that DNA—well, things would have been very different. In August 2012, Alice emailed both of her first cousins and told them she was trying to answer questions raised by a DNA test. In an email to Pete, Alice used the nicknames they reserved for each other—Pete was "Elder," for elder cousin, and she signed off as "Younger." "Just when you think the world is solid," Younger wrote to Elder, "something happens to shake things up." She laid out some of the possibilities she was hoping her cousins' DNA could resolve, including the theory that their shared grandmother had had an affair. Most likely, it was all much ado about nothing. "The test interpretations may need refinement," she wrote.

The year 2012 was not like now—everybody and his brother hadn't taken DNA tests, and plenty of people didn't know what they were. It was not a certainty that the two cousins would agree. But Younger included a link in her email to Elder to some videos explaining DNA testing, and said she and Gerry would pay for the test.

"What do you think?"

Elder wrote back that yes, he would take the test, and so did Alice's maternal first cousin. Now the investigation could begin.

* * *

For a long time, Bennett Greenspan was right about the niche market for consumer genetic tests. The early years were slow going for awareness of what DNA testing for genealogy and ancestry was. It wasn't till about 2013—thirteen years after Greenspan's launch—that consumer genetics companies like his collectively tested their one millionth customer.

FamilyTreeDNA was for years primarily a company for people like Roberta Estes, the silver-haired genealogist, who heard about at-home DNA testing well before most Americans, in the early 2000s. (This was shortly before she managed to track down her long-lost half-brother, Dave, from her father's "other" family, through genealogical research and a private investigator.) Estes was curious about the Y-DNA testing Greenspan was offering for male customers. If she got some male relatives to test, she wondered, could this genetic analysis help her trace the Estes line? She called up Greenspan—he was not only the kind of entrepreneur to call a scientist at home, but also the kind of entrepreneur to get directly on the phone with a customer—and asked him how the science worked. Greenspan recognized another seeker when he met one. He talked to Estes for about ninety minutes, explaining how his Y-DNA test could help her.

Estes was convinced, and she began the first of twenty surname projects, using DNA combined with traditional genealogical methods to understand the lineage of her own family and others. She had worked in technology for years, had an MBA and a master's in computer science, and her mind bent naturally toward data and logic and information systems, much like Alice's. Eventually her skills would prove immensely valuable in a field—genetic genealogy—that she, along with other citizen-scientists, would help create. In time, she also consulted for Greenspan.

When I visited Greenspan, he told me he started his company because of his interest in people and where they came from. "I didn't start this to make a lot of money," he said. But as it happened, FamilyTreeDNA has turned a profit every year since it has existed. It is a small, closely held company of about 150 employees, which Greenspan told me is privately owned by himself and his business partner, Max Blankfeld. They could not have imagined back in 2000 that at-home DNA test kits would blow up, that one day they'd be building their own lab, and then expanding it, and taking on contracts for other genetic testing companies. They could not imagine how busy the lab would become over the holidays, when DNA testing kits become the quintessential holiday gift for the person in your life who already has everything. In the company's first year, when FamilyTreeDNA analyzed a mere three hundred consumer kits, they could not imagine their company's kits would someday be available for sale all over the world.

Greenspan is a talker. He is friendly, occasionally brusque, a born salesman, happiest when he can unfurl a good yarn, and sometimes prickly when he can't control the narrative. His company occupies several floors in a bland brick building off a freeway in Houston. On one wall of his office is an assortment of framed political buttons going back to President William McKinley, and on another, a sheet of white paper at least six feet long shows his family tree going several generations back. When I saw it, I asked Greenspan about the genealogical brick wall that had started his odyssey, and his company. Had the man in Argentina with the unusual surname actually turned out to be a cousin? Greenspan smiled, and walked me through the lengths he'd had to go through to find out, using Y-DNA testing because autosomal DNA testing hadn't yet come along to make genealogists' lives so much easier. He had to send two kits: one to the fellow in Argentina, and one to a male cousin on his mother's side who lived in California. And? Greenspan walked me over to his family tree wall chart and pointed to the Argentinian cousin's name.

For lunch, we drove to a Mexican restaurant where the portions were so large we shared a platter of fajitas. Some weeks before, I had paid for an autosomal test through FamilyTreeDNA, emailing Greenspan to see if he'd be willing to walk me through my results when I was in Texas. I had questions about methodology and definitions, and about why the company's estimate of my ancestral mix differed from the results I'd gotten from 23andMe. Now, as I scooped chicken and onions, Greenspan surprised me by rattling off his company's assessment of my ethnic percentages from memory. He remembered the last name of one of my relative matches (Mendoza), and the number of relative matches I had in his database (eleven thousand), and he recalled even the most granular details of the ancestry report his company had issued for me— I was possibly as much as 2 percent Middle Eastern. I didn't remember seeing much of this, but later when I checked online, I saw he was right about all of it. He'd paid more attention to my results than I had.

When we got back to Greenspan's office I sat by his computer while he signed in and showed me his own DNA results. Examining his relative matches, he knew how certain fourth cousins were related to him through the same set of great-great-great-grandparents. He recalled which distant cousin's father had discovered a relative in Mallorca. "Here's another guy on my dad's side—it's my dad's father's sister's grandson," he said, pointing to the screen.

I probably don't need to say this but: Most people can't do this. Maybe we lack the memory or organizational capacity in our brains. Maybe we simply don't care that much. If America is turning into a nation of seekers—so many of us doing DNA testing on a lark, then backing into family discoveries without quite meaning to—it has taken the passion of longtime genealogists like Greenspan and Roberta Estes and the volunteer researchers at the Family History Library to get us there. They cared about the past before it was easy, before it was sexy and high-tech and quite so commercial, back when it was associated with colonial dames and retirees.

And because many of these original seekers got to genetic genealogy sooner, they learned early on how it can delight—and how it can disappoint.

* * *

For nearly eight years, Roberta Estes had her half-brother, David, in her life. The potato runs, the rescued dogs, the foul mouth, and the steady love. Then David got sick.

Many years before, when David had served as a marine tail gunner in Vietnam, he'd been shot down and received a blood transfusion; almost three decades later, he developed Hepatitis C from that transfusion and then liver cancer.

Roberta was a genealogist, so it made sense that she'd tested David's Y-DNA over the years as part of research into the Estes line. But it wasn't till shortly before David died in 2012 that she ordered autosomal DNA tests for him and for herself, and understood why his Y chromosome didn't match those of other Estes men she'd studied. She and David were not biologically related. Her father—the man who'd had two families—was not actually David's biological father, a fact her father likely never knew himself. David's mother had had a relationship with someone else. Roberta tried to talk to David about what she'd discovered, but he made it clear he didn't want to know the results from the DNA test, so Roberta never told him. The knowledge would have wounded him, and he was sick. And besides, David *was* her brother, in all the important ways.

At David's funeral, the room was filled with his trucker friends, who left their big rigs parked around the funeral home with engines running in a tribute to David. His wife was sobbing. The preacher didn't show. Roberta realized there was no one else to give the eulogy. So she did it.

For years after that, Roberta searched for David's family, wanting to give David a kind of closure, wherever he was. She vowed to find his biological father. Just as she'd learned to decipher patterns of genetic relatedness using Y-DNA, now she learned to decipher autosomal DNA. She was still administrator of David's 23andMe account, so she searched among his matches for close relatives—David's DNA was still working for him years after his body was cremated.

In time, Roberta figured out who David's father had been, and posted about it on her blog in a series titled "Dear Dave." "Dave, meet the man we believe is your father," she wrote, below a grainy black-and-white image of a man with a narrow face and deep-set eyes, who looked exactly like his son. "Dave, meet your half-sister Helen," she wrote, posting pictures of herself with Helen, the two women smiling and hugging.

Funny thing about Roberta and Helen. Because they were both David's sisters—though one woman never had the chance to meet him in life, and the other loved him even though they weren't genetically related—it seemed to Roberta as if they were each other's sisters, too.

When I spoke to Roberta Estes, her depiction of David was so real, I felt like I knew him—his rough edges, his warm heart. She told me that once, when he came by her house for an impromptu visit along a trucking run, he saw some medical paperwork spread out on the table—Roberta thought she might have cancer, though she'd later turn out to be fine.

He confronted her: "Were you going to tell me?" If she was sick, David said, he'd sell his house and come take care of her. She could tell he meant it.

"You can't do that," she said.

"You can't tell me what I can and can't do," her brother replied.

Roberta cried when she told me this story. "I don't know if I've ever had anybody in my life love me like that," she said.

The story of David was not an indictment of DNA testing, Roberta told me. Genealogy had given David to her; DNA could not take him away. The legacy of her father—his brilliance and his dysfunction—had taught her that genetics only goes so far. "Your genes neither guarantee you nor condemn you to anything," she said. And genetics is not the only measure of family. DNA is just DNA—it could not account for her and David.

"It's not love," she said.

* * *

When the 23andMe kits arrived at Alice's house in Vancouver, Washington, and at her sister Gerry's house, in Forest Lake, Minnesota, they spit into the vials, sealed them up and sent them back. Elsewhere in the country, two elderly men—Alice and Gerry's first cousins—were also spitting into vials, just the first of many people who would eventually be drawn into Alice's quest. In 2012 this wasn't as evident to many Americans as it is now, but when one person spits into a vial or swabs her cheek, her whole family is implicated. Because we share DNA with others, one woman's genetic mystery can never be hers alone—which meant that in one way or another, Alice's test results could inevitably prove to have implications for many other people.

For Alice, the waiting was fraught this time, because she felt what she was waiting on, really, was a verdict on herself. If the 23andMe test were to replicate the findings in the AncestryDNA test, it would mean that one of her parents wasn't who he or she was supposed to be. And that would mean Alice wasn't in some sense who she thought she was—a bald fact about herself that Gerry might or might not share, depending on whether they had the same parents. Alice wrestled with this idea: She *was* still herself, of course, but if AncestryDNA was right, it meant some important facts she'd long thought about herself and how she'd come to be were wrong. The context for her life was wrong. And that's a pretty big thing to be confronted with when you've taken a DNA test, as Alice had, "just for fun."

4

YOUR TRUTH OR MINE?

It makes sense that so many adoptees turned to genetic testing when companies like Ancestry unveiled their services. It was adoptees, and the volunteer "search angels" who helped them—a largely female band of citizen-scientists, some of them adoptees themselves—who pioneered many of the techniques that seekers turn to when they search for kin. Within the adoptee community, there is much talk of how deeply frustrating and destabilizing it is to lack basic information about the circumstances of one's own birth because of state laws restricting access to records: the names of one's parents, one's own birth name, important medical information. Not to mention getting answers to other, more profound, questions: Why was I given up? Does my biological mother want to know me? Before the age of genetic testing, an adoptee could spend decades looking for her biological family, and piles of money hiring private investigators, and still get nowhere.

In the pre-genomic era, the prospects for Jacqui Ochoa were bleak. Her longing was palpable. Jacqui wanted things many of us take for granted: She wanted answers to her questions, and she wanted a beginning to her personal narrative, an origin story. Most of us grow up with this story, and it's as familiar to us as the sight of our own face in the mirror: the look of our parents when they were young, as glimpsed in old photos; the story behind how they planned for us (or didn't plan and were happily surprised); their description of how the sky looked and the wind blew on the day our mothers checked into the hospital. How we emerged fat or scrawny, squalling or silent, sick or hungry.

Jacqui wanted her beginning. And she wanted family—perhaps that most of all.

Jacqui Ochoa was a foundling, which is a term for a child abandoned and found, a term that made Jacqui think of a little bird when she first heard it. In October 1965, she was a four-day-old baby tucked into a basket and left in front of a pastor's home in Van Nuys, California. She was adopted, and grew up with four other adopted siblings, and with parents who showed her unconditional love. They made her feel wanted. They told Jacqui her mom must've loved her so much. Jacqui grew up knowing that when she was given away, she'd been bundled in a basket, with a woolen blanket and a soft bunting bag—plus, as she later learned from a news clipping about her own case, diapers, "one undershirt, one pair socks, baby oil and powder."

But despite her upbringing, Jacqui grew up with a feeling there was something wrong, something fundamentally unlovable about her. She described it as a hole in her gut. She thought about the woman who'd given birth to her.

"How could she walk away?" asks Jacqui, who is now a school-teacher in Orange County, California. "That's what I thought about for thirty-five years."

Jacqui wanted answers. She wanted to learn the identity of her mother, and to understand what led her mother to give her up, and to know her own medical history, but her adoption records were sealed. She began searching at fifteen, writing letters to the local county agencies, petitioning the county court. Over the years, she managed to speak to the police officers who were called when she was found, and to read their police report, and to see the house where she was abandoned. As the mother of two boys, she tried to imagine the desperation that would lead a woman to walk to that doorstep and leave that basket, and then walk to a pay phone and call the pastor's home and inform the voice on the other end to go to the front door. She wondered if her mother might be out there, ready to be found but thwarted by the fact that Jacqui did not know how to find her. She placed classifieds in the Los Angeles Times. She searched the Internet when that came along.

When Jacqui first became aware of autosomal genetic testing, she knew it could help her, but the price was too high for a single mom. In the absence of a DNA test, or the unsealing of her adoption records, she needed dumb luck. She needed that one-in-a-thousand, needle-in-a-haystack breakthrough that was the stuff adoptees dreamed of in the pre-genomic age.

Which is what she got.

One day in 2016, almost four decades into her search, Jacqui got a letter from a social worker at the Los Angeles County Department of Children and Family Services. They'd been in contact with her siblings. Would Jacqui like to meet them? Jacqui burst into tears. Yes, by god, after thirty-five years of searching, she'd be thrilled to meet someone from her biological family. This felt like everything, the consummation of so many years of looking and waiting and wanting.

* * *

The landscape of the consumer genomics market now would have been barely recognizable a decade ago. One study by scholar Andelka Phillips, then at the University of Oxford, found that as of January 2016, at least 246 genetics testing companies across the globe were selling their wares directly to customers online. These tests are generally lightly regulated, with the exception now of health-related testing such as that offered by 23andMe, which is more regulated in the United States by the Food and Drug Administration. Not all DNA testing companies offer services related to predicting ethnicity and finding relatives; indeed, the spectrum of services they offer is dizzying, and their usefulness and accuracy sometimes dubious. They range from the paternity tests you can pick up at Walgreens to tests that look specifically for African or Native American ancestry, to others promising DNA-based matchmaking services. Phillips's survey placed consumer DNA tests into a long list of categories that included "child talent," "nutrigenetic," and, most ominously, "surreptitious." The catchiest company names I've seen are "She Cheated" and "Who'zTheDaddy?"

"Have you found an article of clothing with a suspicious stain?" asks the website of one Florida-based company called All About Truth DNA Services, which informs readers that "aprrpoximately [sic] 60%

of husbands and 40% of wives will have an affair at some point," and recommends consumers wait for their "suspicious item" to dry and then send it in for testing. Also accepted: cigarette butts, toothpicks, hair.

While the lion's share of DNA companies cater to questions of ancestry, health, paternity, and relatedness, much of the emerging consumer genomics market falls into lifestyle and fitness categories, encompassing products *The Atlantic*'s Sarah Zhang has likened to horoscopes: "vague, occasionally informative, sometimes amusing." Their claims, and the science used to back them up, are of varying quality. Some tests, met with horror by a wide swath of researchers, promise to offer insight into children's athletic ability. Another company offers an "Inborn Talent Genetic Test" for children, to help with career profiling—the better to maximize "the chances of them becoming an elite in life." Marketing "faux scientific authority," these kinds of tests aren't just harmless entertainment, warns one scholarly paper; they threaten to diminish consumer confidence in the clinical genetic tests that doctors order to guide medical decisions.

The landscape is confusing for the average consumer, and it can be hard to tell what genetic tests to take seriously. Large ancestry-testing companies, like AncestryDNA and 23andMe, may be characterized as "recreational," but they employ teams of scientists and rely on robust data to understand genetic relatedness and to track patterns of ancestral heritage (even if the latter is imperfect and constantly being refined). On the other hand, when 23andMe announced it was teaming up with a health-coaching app, and allowing customers to integrate their genetic results to help generate personalized diet and exercise advice, a number of geneticists were skeptical, concerned the company was getting ahead of research. What is a consumer to believe? A few years ago, Helix, originally a spin-out of genomics giant Illumina (which makes many of the chips and machines used to analyze DNA), unveiled a "DNA app store" allowing third-party companies to sell products off its DNA testing. These have included the Mayo Clinic GeneGuide, a test that requires the sign-off of a physician and, with the help of Mayo Clinic professionals, interprets your genetic material for insights into things like disease risk and carrier screening, but they also included the "Vinome Wine Explorer," which claimed your genetic data could help predict what wine you'd like, a concept one geneticist described as "completely silly." Helix has since announced a shift away from this "consumer-initiated" model, but there is still a lot of confusion over what genetic testing can and should be able to tell us.

One spring day, I found myself watching an ad for a special partnership between 23andMe and Lexus, which promised to find cars optimized to people's genetics. It's a credit to how out-there some DNA testing claims have become that it took me a few seconds to realize this was an April Fool's joke. As ludicrous and playful as that ad turned out to be (the driver licks the steering wheel to start the engine), it hit on a deeper message rooted in a suite of cultural messages we get about our genes. Consider the marketing campaigns that consumer genetics companies actually do run. During the 2018 FIFA Men's World Cup, for which the US team failed to qualify, 23andMe urged people to root for a team "based on your genetic ancestry"—they called this campaign "Root for your Roots." AncestryDNA has partnered with Spotify to create custom playlists based on the ancestral regions that customers hail from. "Solidify a true connection to the motherland," suggests one ad by a company called African Ancestry. "Know who you are"—as if DNA might know us better than we know ourselves, might act as a kind of historical id, reminding us of cultural affinities forgotten over generations but remembered in our cells.

These efforts are targeting—and reinforcing—a deep-seated belief that if we peer closely enough, we'll be able to decipher nearly everything about ourselves, our likes and loves, from the ACGTs along the strands of the double helix of our DNA molecules. It is an idea we've held for decades. In *The DNA Mystique: The Gene as a Cultural Icon*, which was published in 1995, sociologist Dorothy Nelkin and historian M. Susan Lindee warned of the rise of genetic essentialism. Examining portrayals of the gene in mass culture, they found a tendency to point to genes as the explanation for "obesity, criminality, shyness, directional ability, political leanings, and preferred styles of dressing. There are selfish genes, pleasure-seeking genes, violence genes, celebrity genes . . ." In popular portrayal, good genes and bad genes lead to good and bad traits.

We're such believers in genes that a recent Stanford University study found that informing people of their genetic predispositions for certain traits—rather, *mis*informing them, by telling them whether they had certain gene variants associated with exercise capacity and obesity, regardless of their actual results—influenced their actual physiology. Those told they had low-endurance versions of a gene variant did worse on a treadmill test, with poorer endurance and worse lung function (even if they didn't actually have that gene variant). Those told they had a variant that made them feel easily sated felt fuller on average after being given a meal, and tests revealed their bodies had produced more of a hormone that indicates feelings of fullness. By believing they were genetically destined for something, these subjects appear to have made it true.

In their book, Nelkin and Lindee looked back at the eugenics movement earlier in the century and saw thematic links between the 1990s obsession with genetics and those old notions of heredity. Yesterday's "Better Babies" were "still a highly desired reproductive commodity," while yesterday's feebleminded women, forcibly sterilized so as not to pass their problems on to the next generation, had been transformed into contemporary welfare mothers said to be birthing tomorrow's poor and criminal classes. "Ideas about heredity have as much to do with social meaning as they do with scientific research," they wrote.

Beneath all this, the authors argued, lay the mystique of the genes: "DNA has assumed a cultural meaning similar to that of the biblical soul. It has become a sacred entity, a way to explore fundamental questions about human life, to define the essence of human existence." Like the soul, DNA in this reading has a moral meaning and has implications not only for a person's sense of identity but for her place in society. Twenty-five years after that book came out, we still talk about DNA in quasi-religious terms—"the language," to harken back to Bill Clinton, "in which God created life." And perhaps there's something deeply human about this. Cultural psychologist Steven J. Heine has written that "in every society that has been investigated, there is clear evidence to show that we are predisposed to think of the world as emerging from hidden underlying essences"—whether that be blood or chi, humors or souls. Essentialism "is one of the most persistent and widely documented psychological biases."

Heine writes that we tend to think of essences as being "deep down and eternal," as defining and dividing categories within the world—black from white, dog from cat. When it comes to the genetics of traits and disease, Heine says, we are prone to a related outlook he calls "switch thinking"—the tendency to assume that particular genes control these things in a direct and fatalistic fashion. Yet it is not true that there is an "infidelity gene," as some outlets have reported, for instance, that makes a partner's cheating inevitable. "The vast majority of genetic influences on our lives operate in immensely complex ways and are not at all accurately captured by our essentialist biases," Heine writes.

Kristen V. Brown, a Bloomberg journalist who covers the intersection of technology, business, and health, told me she blamed some of this essentialist thinking on the Human Genome Project, "because part of the way that the Human Genome Project was sold to the masses was this idea that your genome explains everything." She added: "And then we decoded most of the important parts and were like, 'Shit, this *still* doesn't explain everything.' . . . But that was the marketing message and it was a good one, and it stuck."

So when companies urge people to root for a soccer team based on genetic heritage, or promise an exercise plan based on their DNA, this homes in on an idea that already holds great currency in the popular imagination. We are eager to know more about ourselves, and within the consumer space, ancestry testing appears to be driving the market for self-discovery. An Ancestry ad during the 2018 holiday season showed Kelly Ripa ordering biscotti in imperfect Italian, since she had just discovered she was "74 percent Italian!" What fun! Who wouldn't want an excuse to expand their diet of baked goods? Seemingly less fraught than health testing, ancestry testing is, as Brown puts it, "the killer app."

The market may change over time, of course. As our understanding of genetics improves, things like pharmacogenomics (the relationship between drugs and genes) and nutrigenomics (the interplay between nutrition and genes) may become much bigger forces. What's now considered "recreational" health testing may become more clinically relevant, and the genetic health market in general may prove to be bigger than that for ancestry and genealogy. But for now, as University College London researcher and genetic genealogist Debbie Kennett points out, the largest genomic dataset in the world isn't in the hands of governments, pharmaceutical companies, or research organizations. "Instead," she writes, "it is the ancestry companies which have been accumulating most of the genetic data."

Today, four companies lead the pack in terms of database size for autosomal tests: AncestryDNA is in the lead, then 23andMe, then the increasingly popular MyHeritage DNA, and, finally, FamilyTreeDNA. The bulk of these tests is being bought in the United States. That they are big business goes without saying, though the different companies operate on different models. Ancestry and MyHeritage have primarily focused on combining DNA testing with subscription-based access to genealogical materials, while 23andMe has carved out the health risk and traits market alongside ancestry testing. Meanwhile, in addition to selling consumer tests, FamilyTreeDNA's lab takes on work for other genetics companies, universities, and hospitals. And there are relative newcomers trying to break into the market.

In the fall of 2018, I attended the New York State Family History Conference, where a big session was led by an up-and-coming UK-based company, Living DNA. One of the brand's ambassadors did a series of cartwheels up the long aisle of the huge ballroom, eliciting gasps from the audience, and the company's co-founder David Nicholson made a strong pitch for the serious genealogists in attendance to buy his company's test. He asked if there was anyone in attendance who *hadn't* yet taken a DNA test; out of the rows upon rows, only four people raised their hands. Yet Nicholson was counting on the fact that he was offering something so unique that not only DNA newbies but even people who'd already tested at two, three, or four other DNA ancestry companies would want to test at his, too. He explained the strengths of the company's forthcoming relative matching system, and showed the granularity of the company's ethnicity estimate, explaining that Living DNA could pinpoint DNA from twenty-one regions in the British Isles alone, even distinguishing between genetic heritage from the county of Devon and that from the neighboring county of Cornwall.

But if Living DNA wanted to catch up to the other companies, especially in the United States, it would have to hustle. It had launched in 2016, only four years after Ancestry started offering autosomal DNA tests, but those four years had been key in the business of DNA testing. Its biggest rival already had more than ten million people in its database by then.

* * *

News outlets outdo each other trying to quantify the hugeness of AncestryDNA's database of genetic data. It's "the world's largest collection of human spittle, numbering in the hundreds of gallons," McClatchy observed a few years back. The approximately 1.5 million kits the company sold between Black Friday and Cyber Monday of 2017 added up to "2,000 gallons of saliva—enough to fill a modest above-ground swimming pool with the genetic history of every person in the city of Philadelphia," wrote *Wired*. The pace of the company's growth is fairly astonishing: It debuted its autosomal DNA test in 2012, several years after 23andMe. Yet when I visited its headquarters in the summer of 2018, it was just about to announce that it had double the number of DNA customers that 23andMe had.

Ancestry as we know it today began with two companies coming together. John Sittner founded a genealogical publishing company called Ancestry in 1983; it produced genealogical reference books, as well as the *Ancestry Newsletter*, which later became a magazine. Meanwhile, in 1990, two Mormon entrepreneurs named Paul B. Allen and Daniel Taggart founded a company called Infobases to sell religious and educational texts, first on floppy disks and then on CD-ROM. In 1995, Allen and Taggart moved into genealogy, licensing some of Sittner's reference books and packaging them with family tree software and other resources into something called the LDS Family History Suite, which made a million dollars in less than half a year. Allen told me later that this was a breakthrough moment for him: To be sure, Mormons were more interested in genealogy than the average American, but still, they made up just a tiny fraction of the US population. What was the potential for the genealogy market if he could broaden his audience?

Realizing they needed a genealogically oriented brand, Allen and Taggart started buying into Sittner's company, eventually coming to own it outright. They launched Ancestry.com in 1996, offering genealogy hobbyists free access to the Social Security death index for their research needs. Allen started surveying online visitors to the site, asking them questions about what they'd spent on various genealogical endeavors over the course of a year. How much on reference books, family tree software, genealogical magazines, travel? What about on postage and photocopies? When he crunched the numbers, he realized his average visitor was spending more than five hundred dollars a year on this hobby. "This is a multibillion-dollar industry and nobody's noticed it yet," Allen realized. "Why don't we digitize everything they're spending money on and make it possible to do it all through one subscription?"

And so they did.

The company dabbled in genetic testing over the years, offering Y-DNA and mitochondrial DNA testing, but these tests had limited appeal. The game-changer was the 2012 debut of autosomal DNA testing, the one that Alice Collins Plebuch tried out in beta, after Ancestry purchased the DNA assets of the nonprofit Sorenson Molecular Genealogy Foundation, and used them to help build its reference panel and make its ethnicity estimates. Suddenly, all of Ancestry's strengths could be brought together into one lucrative venture. The company had gone public in 2009, and in 2012 it was bought by an investor group led by a European private equity firm. At the time, Ancestry was valued at $1.6 billion; when more investors bought stakes in the company four years later, its value had gone up by a billion dollars, according to the *Wall Street Journal*. Since then, it has talked of going public again, and estimates of its value have continued to rise. It generated more than $1 billion in revenue in 2017 alone.

Compared to FamilyTreeDNA, which has an understated presence in a building it shares with other companies, Ancestry's headquarters can't be missed. The headquarters in Lehi, Utah, completed in 2016 at a cost of $35 million, is a gleaming, modern structure—two connected buildings housing about half of the company's 1,600 employees. The lobby windows look out on two majestic mountain ranges, barely visible through a haze on the day I was there. It had been difficult to get permission to tour; access to the company's headquarters was tightly controlled, and visitors had to sign confidentiality agreements just to get past the lobby. Ancestry is almost omnipresent in this area; in addition to the headquarters, its customer service operation is twenty minutes away in Orem, while its team of professional genealogists, who are available for hire, are located forty minutes away in Salt Lake City.

Once I was inside, I was met by Jennifer Utley, the company's director of research and its longest tenured employee, who had been at Ancestry for over twenty years. When she started, the company had just established an Internet presence, and Utley's work was editing genealogy reference books and *Ancestry Magazine*. Now, clocks on omnipresent video screens throughout the headquarters showed the current time in the company's many offices throughout the world, including in San Francisco, Dublin, Munich, and Sydney. About eighty paid interns were eating in the cafeteria, just off the lobby, getting ready to do "speed dating" with executives so that they could learn what it's like to work there. Utley took me upstairs, showing off artwork with a genomic theme: hanging pendant lamps inspired by the double strands of the

DNA helix, and a massive, colorful bar graph representing the fifteen principal biogeographical ancestries of eighty-four human populations. Large portrait photographs of some of the company's employees were mounted on the walls alongside images of their ancestors.

Utley introduced me to two members of the content acquisitions team, who told me they travel all over North America to work with archival facilities, mainly at the state level, digitizing and indexing old records. In exchange for this, the company is given permission to place the records on its website so that its millions of paying subscribers can access them. The company has a tiered subscription model offering access to things like birth, marriage, death, and census records, international and military records, passenger lists, and old newspapers, depending on how much you pay. Based on the subscription you require and the length of time you need it, this could run into several hundred dollars for a year's worth of access. A customer who is simply trying to solve a few basic questions about her great-grandmother could conceivably pay twenty dollars for a month of basic access and never spend any more—assuming she's not converted into a seeker once she gets access to all those astonishing records.

Paul Allen, who left the company in the early 2000s, told me that one way to think about the brilliance of Ancestry's model is to frame its genealogical product as the opposite of breaking news. "The value of the records *increases* over time; they don't decrease over time," he said. "Every year that goes by, more people are entering middle age, where they start to be interested in genealogy, and then they have children and grandchildren." This means the birth record of a single ancestor from 1850, for instance, "is of interest to more people every year just because of aging and population growth."

Utley used her employee pass to unlock a door into a special room into which people were discouraged from bringing pens or open containers of liquid. (An exception was made for my pen and pad.) Ancestry has more than twenty billion genealogical records, and this room was one where the collection grew. There were microfilm and microfiche readers, and machines for scanning. A temporary archive housed the physical objects while they were being digitized—records of yearbooks and city directories, probates and wills. It represented an astonishing effort, but also an endless one—even if it were possible to scan, digitize, and index every document of genealogical interest in the world, human beings keep making more.

"It's not just about capturing images, it's also about preserving them," Utley told me. "We've been in archives where they had to prop things up on toilets" to protect them from water damage. "When records disappear, they're gone forever." The folks at the Family History Library spoke of a similar race against time, mentioning places in the world with so little funding for archives that precious records were kept on dirt floors. It occurred to me throughout my time in Utah that the work of both entities, the corporate one and the church-funded one, must at times feel Sisyphean: trying to save the records of the world before they give way to the ravages of Mother Nature and the forgetting of time. Nearby, a desk held a stack of metal canisters the size of personal pizzas. I glanced at the top one: It contained reels of the *Ottawa Citizen* newspaper, May through October 1972. Who knew what secrets were preserved there, laid down in mundane detail as part of the quotidian duties of putting out a daily paper? Amid all the lines of text in those reels might be an obituary or a marriage announcement containing a single name that was the clue one seeker needed, if only that detail could be ferreted out.

Elsewhere in the building were other operations: marketing, emerging businesses, mobile apps, human resources, and people from the DNA product and features team. Utley showed me a cubicle where several different versions of the company's DNA kit were displayed, customized for the more than thirty countries in which they were being sold. DNA testing at Ancestry works hand in hand with the rest of its business. If you spit into a tube, you'll get your ethnicity estimate and relative matches, but only with a subscription will you gain access to most of your matches' public family trees, or to genealogical records that might help you figure out where your mysterious Italian heritage is coming from. And what's more, sales beget sales.

AncestryDNA becomes a more appealing place to test as its database of customers grows, because people are more likely to find close genetic relatives there.

We walked over to the area housing Utley's own unit, which is responsible for researching and publicizing the genealogical discoveries made possible by the company's resources. (It was Utley's team that uncovered that Emma Watson, who played the witch Hermione in the Harry Potter films, was in real life distantly connected to a woman convicted of witchcraft in 1592, and that Benedict Cumberbatch, who plays Sherlock Holmes on TV, was very-very-slightly related to the author of the Holmes series, Sir Arthur Conan Doyle. If you're willing to extend the definition of "relative" far out enough, the world of genealogy is full of kismet like this.) As we walked, Utley pointed out people who'd worked at the company for a decade or more. She told me she loved her work for the sense of mission it gave her. The company was hosting eighty million family trees on its website; customers' sense of the past informed who they were in the present. She could be selling shampoo or something and that would be fine, but this work gave her the sense she was actually changing people's lives for the better. "There's a real sense of purpose," she said.

At the same time, Ancestry's size and mission can't help but draw scrutiny. A few months before I visited, McClatchy ran a multi-part series on AncestryDNA. One article was headlined "Ancestry wants your spit, your DNA and your trust. Should you give them all three?" The company's entire operation—the DNA testing, the family trees, the incredible treasure trove of genealogical records—is built upon reams and reams of incredibly intimate data—data that tells the stories of people's ancestral backgrounds and genetic traits, not to mention the sacrifices, triumphs, and scandals of their ancestors.

There's a tension inherent in being such a big company managing the genetic information of millions of people. There is a tension between the idea, promulgated by the Sorenson Molecular Genealogy Foundation, that we are all the same, and the idea that our differences can be parsed into pie charts. There's a tension between my right to spit into a vial and have its contents analyzed by a private company—and the right of my brother or mother not to be seen.

* * *

Starting in 1989, Arizona State University researchers joined forces with a small group of Native Americans known as the Havasupai, who were living in a remote area of the Grand Canyon accessible only by foot, horseback, or helicopter. Many Havasupai were afflicted by diabetes, and the researchers hoped to figure out if there was a genetic component to the disease by conducting genetic testing among members of the community.

But many years later, the relationship soured. The Havasupai said they found out in 2003 that the researchers were using their genetic material to look into topics to which they said they didn't consent, including schizophrenia, inbreeding, and the community's ancient population migration. One paper based on the tribe's blood samples suggested the Havasupai's ancestors may have migrated from Asia over the frozen Bering Sea. The Havasupai's origin story was that "it had originated in the canyon and was assigned to be its guardian," as the *New York Times* described it, and this belief was the basis of its sovereign rights, the tribe's vice chairman explained. Therese Markow, the geneticist overseeing the research, maintained that tribal members gave their consent for more far-reaching studies, and that different lines of scientific inquiry can be interrelated—understanding a people's ancestral origins may help in understanding a disease. But in 2010, the Arizona Board of Regents reached a settlement to pay $700,000 to forty-one of the tribe's members and return the blood samples.

This was a decade ago, long before a genetic genealogist and a detective teamed up to use DNA samples in a public genealogy database to solve the Golden State Killer case, and started a new trend for solving long-cold cases, while also driving home that a handful of people can effectively give up the privacy of hundreds or thousands of relatives when they choose to give up their own. It was also before Bennett

Greenspan's FamilyTreeDNA decided to give law enforcement access to its database without telling its customers, angering many genetic genealogists. It was before then-prospective presidential candidate Elizabeth Warren tested herself for Native American ancestry, causing controversy over just what DNA gets to say about Native American identity. It was before Alondra Nelson, Columbia University sociologist and author of *The Social Life of DNA,* observed that genetic testing arguably presents minorities with a Faustian bargain, "where, in order to get information that's been denied because of historical trauma, one has to enter into a conversation with thinking about race in a biological way"—in a way that, throughout history, has often reinforced so many of the racist stereotypes, so much of the reductive thinking, that have oppressed people of color.

And yet the Havasupai case presaged so many of the thorny questions we're only just beginning to think about, including the question of what "informed consent" means. Kim TallBear, a University of Alberta professor who wrote *Native American DNA: Tribal Belonging and the False Promise of Genetic Science,* told me that Native peoples are the "canaries in the coal mine" when it comes to issues of consent in genetic research. Can you truly consent if you can't even imagine what use your genetic material or information might be put to in years to come? What if you can't anticipate how someone else's interpretation of your genetic material will materially impact your sense of identity, your idea of family, or your spiritual beliefs? Bioethicist Thomas May points to the research done suggesting that the Havasupai community had come from across the Bering Sea. "You could take the position, 'Well, that's the truth,'" he said. "You could also take the position that some things aren't open to scientific verification." For science to use someone's body to attempt to disprove something sacred to that person—is that the uncovering of truth or a violation?

TallBear says that reducing the Havasupai story to a clash between science and religion doesn't capture this dispute, nor is it accurate to say that tribes like the Havasupai are anti-science. Origin narratives speak to how Native peoples see the world and their place within it. These narratives are important because they are guides to living, offering moral lessons that inform how Native peoples ethically care for the land and for one another. Some non-Natives "insist on viewing indigenous origin narratives as literal, and if we can't prove them, they're false," TallBear told me. "There are different kinds of truths in the narratives we tell."

The rise of consumer genomics poses questions about the emphasis we put on genetic identity, and what we do when DNA test results come into conflict with the narratives we've long believed about ourselves. How much of your sense of yourself should scientists and algorithms be allowed to dictate? We all hold certain truths to be sacred—they may be truths about where our people came from or about a sibling we love. What we do with genetic information, how we incorporate it into our story, is the tricky part—the part that Alice and Roberta and so many of us are just now figuring out.

It seemed like Jacqui Ochoa, the foundling abandoned on a pastor's doorstep in 1965, had the good luck to sidestep the problems of genetic testing coming into conflict with identity. She didn't need to depend on complex techniques to trace her way to her family, nor did she need to ask herself questions about whether her siblings would want to be found. Instead, she had Vicki, Eric, and Katrina, the siblings who'd been looking for her. As children, after their mother's suicide, Vicki and Eric had learned that their mother had had a little girl they'd never known, and Vicki came across an old news clipping of an abandoned baby she believed might be that little girl. Years later, Vicki and Eric and their half-sister, Katrina, filed paperwork to find her. Staff at the Los Angeles County Department of Children and Family Services matched up the two files—that of Jacqui the foundling and of the three siblings looking for their long-lost sister—according to articles from a few years ago covering their tearful reunion. "It's quite unbelievable that we were able to do it," a social worker for the county agency told

ABCNews.com, describing the bringing of these siblings together as a "miracle."

"We reunited and met and fell in love," Jacqui told me. She and Vicki Costa bonded quickly, trying to make up for lost years, with Jacqui flying to Central California to see Vicki, and Vicki coming south to Orange County. They drank wine and talked till the sun came up. Vicki became an aunt to Jacqui's sons. And while in matters of politics and religion they differ, in other ways they are strikingly similar—both are bighearted, emotionally generous, quick to laugh and to cry. There were times, talking to one woman, when I felt as if I were talking to the other. When Jacqui learned of her mother's emotional intensity, she saw herself in those descriptions, and when she learned her mother had committed suicide, she wondered what this meant for her. Could that instability come for her someday, like a kind of genetic curse?

After a year, it occurred to Jacqui that it made sense to do DNA testing. She was completely certain that Vicki was her sister, yet she wanted the confidence and certainty that only science could give her. It was then, when their lives were woven together, that she discovered Vicki and Eric and Katrina were *not* her siblings. (The Los Angeles County Department of Children and Family Services clammed up, telling me that without a court order, it couldn't comment on the specifics of how it matched up two unrelated families.) Jacqui was heartbroken. She still loved them, and they her, and she remains so close to Vicki that the two still call each other "sister." But Jacqui felt she no longer belonged to Vicki and her siblings in quite the same way. She envied them their genetic bond.

"I felt like somebody just stuck me on a raft and pushed me out to sea," Jacqui said. The final blow came when, with the help of CeCe Moore, the genetic genealogist Alice had also stumbled across, Jacqui located her actual biological siblings, five half-sisters, none of whom embraced her. For the half-siblings on her father's side, Jacqui is an unwelcome reminder of a man she is told was neglectful and mentally unstable. On her mother's side, Jacqui's half-sisters cannot believe their mother would abandon a baby.

"Our mother wouldn't do that," Jacqui says they told her, her voice rising and breaking when she repeated their words. She could never shake the sense of having been abandoned when she was born; now, she felt, it was happening again. Jacqui says her maternal half-sisters told her they didn't trust the DNA results, even though the science of relative-matching is quite reliable. She struggled to make sense of their rejection. How could she find such likeness and warmth from genetic strangers, and yet her own blood cast her out? How could human beings be so wonderful and so cruel? "I just can't believe it," Jacqui said, "that all of the time and the effort and the love and the excitement and the joy and the anguish and the fear and the frustration ended up like this. *You don't believe the science?!*"

I tried to imagine what it must be like for Jacqui's maternal half-sisters to be told that their late mother had given up a baby decades before. Was it impossible to incorporate this into their memories of the woman they knew? Did this unwelcome intrusion invite all sorts of revisions of the past and reassessments of her character? Did it cause them to wonder about the limits of her love? Such a fact—the fact of Jacqui's existence —must pose a cognitive disconnect, like being told the sky is red. This stranger's truth was in conflict with theirs, so her truth had to be wrong.

* * *

The first inkling Bennett Greenspan got that this DNA stuff could be life-changing, and not always in a good way, was before he started selling DNA kits. He was doing what he calls "proof of concept" Y-DNA testing within a large group of men he knew. Even within that early cohort, he found a major and unwelcome surprise, the details of which he didn't want me to publish. "Do no harm" was his philosophy, he told me. But in the very near future, he said, the concept of an adoptee who didn't know her biological family simply would not exist, no matter what state laws dictate.

When we sat in his office under his huge family chart, he rumi-
nated on how the legitimate interests of an adoptee seeking health information could bump up against the legitimate interests of a biological
mother who'd given her child away decades before. Perhaps this birth
mother had never told anyone about her child, and her marriage was
built in part on this omission. Perhaps she never took a genetic test, but
her sister did, and that was all it took.

"How did I know in twenty years they were going to come up with
a DNA test that was somehow going to allow you to find me?" Greenspan asked, channeling this birth mother. And on the other hand, he
pondered, was any of this the daughter's fault? Was she not entitled to
know what and who she was made of? Whose rights should predomi-
nate? "I'm not an ethicist," he said. "Look, I think everything we do
should be guided on, let's try not to hurt anybody. OK? But eventually,
probably, somebody's going to get hurt. Either you as the mother or me
as the kid is going to get hurt when I'm happy to see you, you're not
happy to see me."

Greenspan has a slightly mordant sense of humor. He's started calling the phrase "anonymous sperm donor" an oxymoron akin to "jumbo
shrimp." He told me that when he gives talks, he likes to tell audiences
that it is the duty of grandparents to send their grandsons beer money
so that today's young men won't resort to selling their sperm. Otherwise, he can, like a fortune-teller, predict their futures. In twenty years,
those boys will be getting a knock on the door from someone calling
them "Dad."

5

NON-PATERNITY EVENTS

Jason is in his late forties. He lives in the Midwest. If it were up to him, his full name would be printed here; he is done with secrets. But this is his mother's story, too.

Jason was mostly raised by his grandparents, with his mom and a stepfather in and out of his life. His mom wouldn't talk about who his father was, not when she came to his high school graduation, nor when he was getting married. He was too young to understand, she'd say, or now wasn't a good time.

Jason was in his thirties, a father with two young kids, when he decided to ask again. A relative had recently died, and it occurred to him his mother might pass away without ever revealing the mystery of his paternity. In the pre-genomic age, he was at the mercy of what she and any other secret-keepers were willing to tell him. So, he wrote his mom a letter and put some teeth to his request: He told her that if he did not hear back, he'd start asking around; he'd heard some cousins might know some things. It worked. His mother wrote back with a name and little else. "Here is what you wanted, sorry it took so long," she wrote. "I would just as soon leave as is."

But Jason could not leave as is. He wanted to know his father. He was not only acting out of a desire to know about his dad's medical history, even if that's what he told himself at the time. What he wanted, he knows now, was something deeper. He felt a hole inside himself. "Like a yearning," he says.

He wrote a letter to the guy, who lived a few hours away, and they forged a restrained father-son relationship. The man told him he'd dated Jason's mother in college, and he knew she'd gotten pregnant, but she'd told him the baby wasn't his. The man was kind and polite, but he did not welcome Jason into his life, nor tell his adult children that Jason existed. Instead, once or twice a year, they went golfing together. Once or twice a year for over a decade.

Jason told me later that this arm's-length relationship was what drove him to do a DNA test, in hopes that scientific proof would improve their relationship. You might already know where this is going, but Jason didn't. He never dreamed the man wasn't his father; he only wanted to stop feeling like a secret. So, in 2016, at the urging of a genealogist friend, Jason bought an AncestryDNA test.

That's how Jason discovered he had the wrong man. His purported father wasn't in the database, but that didn't matter. Jason's friend came over and, looking at Jason's list of genetic relatives, swiftly traced Jason into a totally different family, as the likely son of one of three brothers.

In my reporting, I encountered a number of stories like Jason's, stories in which mothers would not or could not talk to their children about how they'd come into the world. Perhaps they did not know who the father was—and how could they tell their children that? Perhaps the circumstances surrounding this conception had been traumatic, had involved coercion or violence. The questions brought back shame or anger or an experience too private to share. They brought back a world in which a pregnant teenager was whispered about and shunned. DNA testing brings old taboos to the fore, secrets that are like the proverbial snowball rolling down the hill, getting bigger and heavier with each passing day. How do you undo a secret?

Only later did Jason look back and see how, in his long relationship with the man who was not his father, he'd been forcing the facts to fit a narrative. He had searched for resemblances between them, and seized on the fact that they were both the same height, both even-tempered and reserved. It was a testament, he realized later, to how badly he wanted this kinship. "You're going to find commonalities with anybody," he says. In one sense, then, Jason's account is a rebuttal to those who believe that DNA testing emphasizes genetic ties over other family bonds. After all, Jason found value in what he thought was a biological bond long before he could prove it with a test. Consumer genomics simply told him he was wrong.

When Jason handed the man a letter, at their next golfing session, twelve years into the relationship, he watched his would-be father's face move through shock and confusion. The man looked up. He told Jason that he was relieved. For twelve years he'd carried the guilt of believing he'd unknowingly abandoned a child. Now he knew he had not. And he said he was sad—"Anyone would be proud to have you as a kid," he told Jason, who cried when he repeated these words to me. The man's words were balm to the sense of shame Jason had felt since he was a child.

Jason went home and once again wrote a letter, with three copies for three brothers, informing them that one of them was his father. The letter had ripple effects—one brother said his wife had learned about it, and it was causing problems in his marriage—but it turned out to be a different brother who claimed Jason. When, at last, Jason met this man at a restaurant, he did not need to search for resemblances. The man's eyes were "like looking into a mirror." The man was warm, gregarious, funny. They sat and talked for hours. The man had asked Jason for an old photo of his mom, and recalled knowing her briefly. He told Jason he already knew they were father and son, though he agreed to do the paternity test Jason picked up at Walgreens. They sat in an SUV outside a post office and swabbed their cheeks together, then sent it off. Within a week, the test confirmed their relationship. The man welcomed Jason into his own blended family, in which no distinction was made between full and half- and step-siblings. Suddenly, Jason had a whole brood of brothers and sisters, several of whom lived nearby, and they all got in touch to tell him how happy they were to meet him.

That first Christmas, Jason's new family insisted he and his wife and kids join them, and it was the beginning of holidays together, and summer weekends, and talking on the phone, and visits to see one another's kids' in their plays and recitals. Jason's wife told me Jason had changed in the years since finding his father. He had become a more confident person. The hole was filled. "My wife said it was a dream come true, and it was—being welcomed," Jason said, thinking back to that first Christmas and starting to cry again. "And nobody was ashamed, and everybody was happy, and it was a great thing."

I could not help but think, hearing this story and thinking of the foundling Jacqui Ochoa, who wanted but did not get the bounty that Jason got, that spitting into a vial in search of family was like spinning a roulette wheel, with no ability to predict the outcome in advance, and the highest of stakes.

* * *

When a person discovers his dad is not his genetic father, this is often known in the world of seekers as a "non-paternity event," or NPE. Because of the popularity of home DNA testing, the term has trickled down from the scholarly confines of genetic genealogy to become an increasingly mainstream acronym. In Facebook groups devoted to finding family through DNA, people are constantly posting about the discovery of NPEs in their families. A nonprofit called NPE Friends Fellowship, dedicated to support and education around this phenomenon, has so far held a conference and a cruise, and offers grant programs, including one to supplement travel costs for those who want to meet their newly discovered biological families. Sometimes, the term "NPE" becomes a way to refer not to the fact of unexpected paternity but to the adult child uncovering it. "I'm an NPE," seekers write, and the term is transformed into a modern and reductive form of identity.

At genealogy events and in academic papers, I came across many other terms to describe NPEs, including "misattributed paternity," "misidentified paternity," "false paternity," "misaligned paternity," "extra paternity event," "surname discontinuity event," and "non-patrilineal transmission." There's the old-fashioned term "cuckoldry," which has a very *Jerry Springer Show* feel, and the rather opaque "extra-pair paternity," a term I found in biology papers on such populations as tree swallows, white-handed gibbons, and human beings. Whatever you call it, questions of paternity are at the heart of most of the DNA surprises America is experiencing right now. "We get a lot of people who found out their dad isn't their dad," says Yoav Naveh, the director of MyHeritage's

DNA team. Some seekers use "NPE" to mean "non-*parental* event" or "not parent expected," to refer to the broader category of cases in which both or either parent is not who the seeker thought.

How common are NPEs? It's hard to say. There is a common misconception that the phenomenon of what one academic study calls "cuckolded fathers" may be as high as 30 percent. But the most reliable recent studies, parsing the question in a number of ways, tend to put the rate in the low single digits, probably between 1 and 2 percent per generation.

More broadly, it is difficult to quantify just how often recreational testers discover significant unexpected things, whether that be instances of misattributed parentage, or surprise half-siblings, or the truth of an ethnic identity hidden from them for years. There are so many ways these surprises can play out. There are cases where adoptees were never told they were adopted, and found out via DNA test. (This is more common among older testers, whose childhoods harken back to a time when more adoptions were undisclosed to children.) There are cases in which people did not know they were conceived by sperm donors, particularly going decades back to a time when assisted fertility was kept secret, and when some doctors were in the practice of mixing a husband's sperm with that of the donor, or encouraging couples to have sex around the time of insemination so that husband and wife could say they never truly knew the genetic paternity of their child. In my reporting, I encountered a number of testers of significant African American ancestry who did not know of this part of their heritage till they tested, for a variety of reasons. Growing up, they were told they were Italian or Native American.

The shared nature of genetic material makes it possible to discover non-paternity events even when a parent and child haven't both tested to compare their DNA. For example, a man doesn't need to be in a DNA database to discover he's not related to a child he thought was his, or that he *is* related to a child he didn't know of; it might be that that child tests and doesn't match a paternal cousin, and things unravel from there. Or perhaps two siblings test, and they discover they share only 25 percent of their genetic material, instead of 50 percent, suggesting one has a different parent.

Ancestry and 23andMe told me they don't gather statistics on how often their customers get unexpected results about their immediate families. Anecdotally, though, Jennifer Utley of Ancestry said that unexpected paternity "happens more than people would expect," though some of this may be because seekers are self-selecting—they choose to test because they have questions. "I think it's impossible to quantify because I see it so much," says CeCe Moore, who, in the years since Alice saw her name on a message board, has become well-known outside the close-knit world of genetic genealogy. Even with a very narrow definition of unexpected findings—if we confine it to consumers who discover one of their parents isn't a genetic parent, or that they have a previously unknown sibling or half-sibling—the number is in the low single digits at least, Moore told me. (Rebekah Drumsta, public relations director for NPE Friends Fellowship, told me she thought the number of testers affected could be as high as 15 percent.) But if you assume just 3 percent of consumers, which is a conservative estimate, spread across thirty million testers in databases, you're talking close to a million people—and more every day. Population geneticist Razib Khan, who has studied the steep slope of consumer spit-kit sales, co-authored a 2018 paper projecting that a hundred million people would be genotyped by 2021, and told me he believed the majority of the American population would have its DNA analyzed by the end of 2023.

Yet 3 percent doesn't even capture the scope of the phenomenon. It doesn't account for the people who haven't tested and may never wish to—but will find out the truth anyway, as a result of a decision someone else made. It doesn't account for the fact that NPE rates increase cumulatively the more generations back you go, which means you may find a surprise in your grandparents' generation if not in your parents'. And it doesn't account for all the people affected by one genetic revelation. Because of the nature of family, each revelation must refract across a series of different people, and in different ways. A man who discovers that an affair decades back yielded a child may also mean his wife discovering this, too, once that child starts contacting him and his other children. A tester who finds out the man who raised him isn't his genetic father is also finding out that the brother he was raised with is genetically a half-brother, a piece of news he may now share with that sibling, while possibly discovering half-siblings he's never met through his genetic father. Every adult who discovers she was donor-conceived may have five or ten or more previously unknown siblings, some of whom will discover this truth in years to come. In a family, everyone is implicated by a secret.

"You see an intact family and you just assume everyone is genetically related, and that's just not always the case," Moore told me. "At first I thought it would just be a trickle—but it turns out to be more like a flood."

Moore has been a consultant for many seasons on the PBS series *Finding Your Roots*, a genealogy show that incorporates genetics to tell celebrities about their ancestors and familial connections. Her background is in television, and she has an intelligence and camera-ready charisma that make her good at explaining not only the technical intricacies of genetic genealogy techniques but also the emotional implications at work in the cases she's seen. I attended a talk she gave in which she explained that she used to try to help eager genealogists in their efforts to persuade reluctant relatives to undergo DNA testing. She understood their enthusiasm—a particular aunt's genetic data might truly be the key to tracing an ancestral line back to colonial America. But, Moore told the audience, she no longer honors these requests, because of the possibility that something else will be discovered.

"People ask all the time at conferences, 'How do I convince my relatives to test?' I say, 'Don't,'" Moore told me. This was both an ethical concern and a pragmatic one, she said; she did not want genetic genealogy tainted by bad press. "If we pressure people who don't want to test and they find out some unexpected surprise, that's not going to go well for our field," she said.

In many ways, the genomic testing revolution represents the end of family secrets. There is a vast army of American men and women searching for their families, and many of them didn't even know they needed to search until they took a DNA test. Was Alice one of these people? Was Jim Collins not the man who'd contributed half of her genetic material, even if in countless ways big and small, he'd made her who she was? And if so, what would she do with that information?

FamilyTreeDNA's Bennett Greenspan told me he tells people: "If you really don't want to know the answer to the question, don't ask the question."

But I'm not sure that's fair. Because we don't even know we're asking it.

* * *

DNA testing companies don't always do the best job warning customers about the unexpected things they may find if they test. A few years back, researchers at the University of Leuven in Belgium looked at the English-language websites of forty-three direct-to-consumer DNA testing companies and found that few warned consumers about the possibility of discovering so-called non-paternity events. But even the warnings that are in place may not be up to the task of getting consumers to pause before spitting and consider what's truly an existential question. To understand why, it helps to consider 23andMe. Among the major companies, it is fairly good at making explicit that its customers may discover things they weren't planning to find out.

"Though uncommon," customers are told before they access 23andMe's "DNA Relatives" feature, "unexpected relationships may be identified that could affect you and your family." The company offers additional warnings on a before-you-buy page on its website and in its terms of service. Its repeated warnings may be due in part to the company's learning curve on health-related results and the transparency imposed on it by the FDA. The company had been selling its genetic testing service for health for several years when the FDA ordered it to stop in late 2013, calling the company's "Saliva Collection Kit and Personal Genome Service" a medical device that required approval. The

FDA said it worried, in part, about false positives for things like genetic risk of breast cancer leading to unnecessary surgery. After its health testing was shut down, 23andMe continued to offer ancestry information and worked with the FDA on how to better deliver its health risk results. In 2015, it gained permission to tell customers about genetic variants for certain diseases, like cystic fibrosis, that they might pass on to their children, and in subsequent years, it was given permission to add genetic health risk tests for diseases including late-onset Alzheimer's, Parkinson's, celiac disease, and three variants in the *BRCA1* and *BRCA2* genes that substantially increase the risk of breast and ovarian cancer in women. In addition to proving its accuracy, the company also had to demonstrate that it was helping its customers understand what their health results meant and what next steps they should take. That meant using language that "your everyday average consumer would be able to understand," says Stacey Detweiler, a medical affairs associate for 23andMe, plus a lot of repetition, and this consciousness about helping customers handle health risk news has to some extent carried over to how it frames the possibility of familial revelations.

Also, 23andMe had a brush with the bad publicity such revelations can cause early on. In 2014, *Vox* published a package of stories about the perils of at-home DNA testing, recounting the story of a certain "George Doe" whose parents split after a 23andMe revelation. Shortly afterward, 23andMe co-founder Anne Wojcicki reversed a decision to automatically opt existing users into seeing their close genetic relatives, and announced that she was hiring a chief privacy officer.

Still, warnings may be of limited utility given how we're engaging with DNA testing. It's hard to fathom life-changing results when you're testing for fun and entertainment. As Benjamin Berkman, a bioethicist at the National Institutes of Health who specializes in genetics and emerging technologies, points out, those undergoing recreational testing are not prepared to have their identities upended. "If you go get a medical test, you know to be worried," Berkman told me. "You don't necessarily know to be worried about the uncertainty of your family relationships." Or, as geneticist Joe Pickrell, CEO of a DNA testing company called Gencove, puts it, "The problem is, nobody thinks it's going to be them."

Over and over, I found this to be true in interviews: Most people think they already know what there is to know about their families. If they have an inkling that the vial they're about to spit into could reveal family surprises for some people, they don't consider this something they, in particular, need to worry about. Even Jennifer Utley, who spends her days at Ancestry steeped in the complications of other people's families—and whose company cautions that testers may "discover unexpected relatives, which could surprise you and your family"—did not think she'd be a statistic. Until the day she was stunned to discover, through her list of genetic relatives, that she had a previously unknown first cousin—a child her uncle and his girlfriend had given up for adoption decades before.

This can have implications beyond how we conceive of family, identity, and trust, though those can be profound enough. "I never ever expected this to happen—ever," geneticist Ricki Lewis said after discovering through an AncestryDNA kit that she had a half-sister, which led her to unravel that her late father was not biologically related to her and that she had been conceived by a sperm donor. In late 2017 Lewis, a genetic counselor and longtime science writer who authors a weekly blog called *DNA Science,* had written an entry headlined "DNA Testing Kits as Holiday Gifts Can Bring Surprises," never dreaming that less than a year later, the surprise would be hers. The revelation came at a crucial time, just as Lewis was deciding how to manage a precancerous condition after previous brushes with the disease. When the half-sister told Lewis about her own past cancer, Lewis decided against a wait-and-see approach and chose to have a second mastectomy. She believes finding out this news when she did may have saved her life.

In the worlds of genetics research and clinical testing, participants are typically given the choice of whether or not to opt in to what are known as "incidental findings": genetic information that's potentially medically important but falls outside of the original purpose of the test. Over time, debate has grown within the medical community over whether to ask patients if they want to be informed of incidental findings—and whether to inform them of such findings when they've asked to opt out of such news. The problem of incidental findings is an important one, in the context of both scientific research and "recreational" testing: If you can't anticipate what you might find out, how can you make an educated decision about what news you'll want to hear?

Thomas H. Murray, president emeritus of a bioethics research institute called the Hastings Center, told me he first encountered these issues over thirty years ago, when he was recruited to create a bioethics center at Case Western Reserve University. He took a call from medical professionals faced with a dilemma: They had a male patient with failing kidneys, and had tested both of the patient's daughters to see if they'd be suitable candidates for a transplant organ. Testing had revealed that one of the daughters wasn't genetically related to her father. Were the doctors obliged to tell her the truth of her paternity?

After much thought, Murray told them, *No.* "All you need to say is that you're not a good antigen match." What the doctors did need, Murray told them, was a policy for dealing with these situations, under which would-be donors were informed that the information they'd receive would be limited to whether or not they were good donor candidates. The doctors would act as gatekeepers, protecting the families from information that was not germane to the matter at hand and could be deeply destructive.

But within the direct-to-consumer world, of course, there are no gatekeepers debating the value of an incidental finding that doesn't feel incidental at all to a woman who just discovered she's the product of an NPE. Is more information always better? Is it always empowering? Information carries its own burden, which is that once you know something, you have to figure out what to do with it. This is true for health-related genetic information: Knowing you may be at elevated risk of developing Alzheimer's is living in an uneasy truce with time. Do you tell your children, or will the information burden them? And this also holds true for the kinds of familial revelations that come from ancestry testing, which is making us all into lonely bioethicists, forced into difficult decisions without the benefit of task forces or review boards. What do you do when you find out you have a paternal half-brother, with all that this implies about your parents' marriage, your father's loyalties, your own identity? What do you do when faced with the prospect of informing your elderly mother that her genetic father was not the man she knew, but a sperm donor?

"We are empowering consumers, patients, family members, caregivers to obtain all sorts of information that we couldn't provide before, and we're just saying, 'Here are the keys to this and go for it,'" says genetic privacy scholar Mark Rothstein. "There's a principle that we teach our first-year medical students: Don't run a test unless you know what you're going to do with the results one way or the other," he says.

We are blundering into these discoveries, assuming that the results of testing will line up with what we already know about our families. Engaging in recreational genetic testing is framed as an act of self-discovery, but it's often assumed those discoveries will come with layers of historical and emotional distance. Consumers I spoke with were often motivated by the desire to discover something cool and innocuous. The fact that DNA kits are given at the holidays and for Father's Day sets the context for how they are viewed: as gifts, the cost of which can be measured in mere dollars.

The major DNA testing companies know this; they style marketing campaigns around the major holidays, with prices dropping as Thanksgiving approaches. "It's the perfect gift for every person," 74-percent-Italian Kelly Ripa promised on Ancestry's home page in November 2018, advertising DNA kits for $59.99. The ad copy promised that "your loved ones" will "have new stories to share—and new ways to grow closer over the holidays." A few days later I got an email from the company: The kit was now $49, its "lowest price ever." In 2016 and 2017, according to Kantar Media, Ancestry spent more than $100 million in advertising each year, dwarfing its competitors, though 23andMe has clearly been trying to catch up. It went from spending $12 million in 2016 to $40 million in 2017 to $50 million in the first half of 2018 alone.

But what about the times when this gift does not help a family "grow closer"? When I wrote about this phenomenon a few years ago for the *Washington Post,* I was struck by a line from novelist Margaret Atwood: All new technologies have a good side, a bad side, and a "stupid side you hadn't considered." Now, consumers are forced to grapple with the unforeseen implications of their curiosity, their generosity. Now they call the company that tested them, confused about why Ancestry is telling them some strange man is their father.

A spokesperson for Ancestry told me it has a team of "member services representatives" to help customers with questions about their DNA results, including "a small, dedicated group of highly experienced representatives who speak to customers with more sensitive queries." The specialized team at 23andMe undergoes months of training for phone calls that require not only scientific knowledge but a great deal of empathetic listening, and the company even launched a special page to help people navigate "unexpected relationships," complete with links for finding therapists. Greenspan told me people who contact his customer service representatives at FamilyTreeDNA with unexpected results often presume the lab is in error. His reps will tell customers that the company has a policy: It'll test the person's sample again for half the price, and if the lab is wrong, the company will refund that money.

And his reps will tell callers that the lab is almost never wrong.

* * *

Even if you didn't mean to ask the question, once it's asked, it will be answered. And once it's answered—well, for many people, there's something pretty compelling about knowing there's a mystery man out there who gave you half your genetic material. How do you not open that box? How do you not want to see your face in his, or to hear the timbre of his voice? How do you not wonder: Would he like you? Would he be glad you came into his life? This is how seekers are made: One question leads to another.

Laurie Pratt inherited the hobby of genealogy from her grand-mother; as a child she used to go on excursions with her to look at headstones. As an adult, she spent many years tracing her own family's history, following her mother's colonial American ancestry and her father's French Canadian roots, before the price of DNA testing dropped enough for her to test herself and her dad, whose tree she was working on.

When Laurie's father's results came in, he did not show up as a genetic relative. Laurie called Ancestry and got a customer service representative who tried to reassure her. "I'm sure he'll show up," she remembers the rep saying. But he never appeared among Laurie's relative matches. So Laurie contacted CeCe Moore, whose name appeared online for her just as it had for Alice—as a kind of lifeline for early adopters of this technology who could not make sense of what a company was telling them. Moore suggested Laurie try testing herself and her father again at 23andMe. Laurie made up a reason why she needed to test her dad again, and again, they did not match. Moore was gentle. She guided Laurie to 23andMe's chromosome browser, showing her how the company had come to this conclusion: Laurie and her father did not have any overlapping genetic segments.

Laurie is in her fifties. She lives in Orange County, California, and works as a ground operations supervisor for an airline. Both of her parents have since passed away, but her mother was "my best friend, hilarious, amazing, always owned her stuff," she says. She went to her mother, who at first said the DNA results were "impossible." But over time, her mom's accounting of whether she could have been with another man changed to "not that I remember."

" 'Not that you remember?' " Laurie repeated, gently teasing her mom. "You sound like Reagan at Iran-Contra."

The first time I spoke with Laurie, she told me she believed her mom honestly didn't remember something that had happened half a century earlier. Later, as she thought back on tiny clues as so many seekers have done, she would conclude that her mother probably did remember; she simply didn't want to have this reckoning with her daughter. But eventually, Laurie's mom recalled some bare facts about a short-lived relationship during a period when she and her husband were briefly separated. Her mother had two requests: "Don't tell your siblings until after I'm gone," and "Don't tell your father." That was fine by Laurie, who said she never would have told her father anyway. "This is the guy who did the job," she says. "I can't imagine calling anybody else 'Dad.'"

Still, the genealogist in her was deeply curious. She threw herself into figuring out who her genetic father was, and it took years. Using genetic genealogy techniques, she traced the family histories of cousins whose DNA she matched online. She combed obituaries for men who were likely candidates. Sometimes she texted her mom photographs of the men—*Is this the one? What about this guy?*—and Laurie's mother pointed out that the fellow she'd slept with hadn't looked like some old guy from an obituary. He'd been a young man.

At one point, Laurie's search became a kind of dark comedy of misattributed paternity. She enlisted the help of a third cousin she'd found on her relative match list, who agreed to test her own parents in hopes of helping Laurie—and in the process this cousin stumbled across the fact that her *own* father was not her genetic father. Even as Laurie searched, her own parents—who had divorced when Laurie was an adult, though they remained close—died within a year of each other and were buried next to each other.

For cases of unknown parentage, much of the process of searching for relatives through DNA databases relies on the principle of triangulation. For years, the technique was premised on something called segment triangulation, which relies on specific segments of identical DNA shared among relatives; in more recent years, CeCe Moore and Laurie Pratt and others have come to rely on a technique called pedigree triangulation. The exact techniques vary depending on a number of factors, but these days, Laurie told me, the basic idea goes like this for a seeker—let's say an adoptee trying to figure out the identity of her birth mother—using the biggest database, AncestryDNA. The adoptee looks at her closest relative matches and examines the amount of DNA she shares with them, as well as cousins they share in common and any family trees that might be available, to figure out how she might be related to them, attempting to isolate her maternal matches from her paternal ones. She might turn to the Internet to figure out the full names and geographic locations of these relatives, if she can.

Then, using genealogical records and tools available online at Ancestry, FamilySearch, and elsewhere, including obituary records, city directories, school yearbooks, and places like Newspapers.com, and drawing on existing trees, social media accounts, ancestral hometowns, common surnames, and shared ethnicity, she builds family trees for her maternal DNA cousins and goes back through time looking for the ancestor she shares with those cousins. (In genetic genealogy forums, seekers refer to this as MRCA, which stands for "most recent common ancestor.") Then, the adoptee builds the trees forward through time—something Moore calls "reverse genealogy"—figuring out what marriages took place, and who are the offspring of those marriages, to isolate which branch likely contains her maternal grandparents. If she's right, the DNA and family trees of her other relative matches should line up with her theory. Once she knows her grandparents, she can very possibly divine her mother, or at least one of several sisters as a candidate for her mother. Then, it's a matter of figuring out what to do with that knowledge, which can be its own kind of journey.

These days, Laurie told me, she thinks she could solve the question of her own paternity within a couple of months, because the databases are so much bigger. But for years, her matches were too distant for her to make headway, and many of her paternal DNA cousins hailed from a small town in Maine with families that had intermarried, sharing last names and genetic material in such a way as to make the job extremely difficult. She became one of the volunteer "search angels" in Moore's large DNA Detectives Facebook group, and when her own search stalled, she solved other people's cases.

Eventually, in early 2017, after researching and creating 118 family trees for DNA cousins in hopes of understanding where she fit into them, and building back in time to the late 1700s, Laurie traced her way to the man who appeared to be her genetic father. She scoped him out online, learned about his religious affiliation, his charity work, and his family's backstory, and then sent him a letter. A few days later, by happenstance, the man showed up as her father at Ancestry, having just tested his saliva. Perhaps he'd been given the kit as a gift from his daughter for Christmas, Laurie thought, since the man's daughter was listed as the administrator on his DNA account. If so, it would prove to be an ambiguous gift.

The man called Laurie at the number she'd provided in her letter, and it appeared he hadn't seen his matches yet, because he suggested that perhaps his brother was Laurie's father. "Your Ancestry results came in today and it came back that you're my dad," Laurie told him. "I didn't test for that," the man replied—because, after all, he hadn't. He didn't know he was asking *that* question. "Did you test in a tube recently?" Laurie asked him, trying to thread the gap for the man, to explain that he had, in fact, tested for his genetic relatives, even if all he'd wanted was one of those cool pie charts that show you how Irish you are.

They talked for a little while. The man did not remember Laurie's mother. "That's OK," Laurie told him, "she didn't remember you either." He was a religious Catholic and said he felt terribly guilty about the idea that he'd had a daughter he didn't know about, though she'd been conceived before he was married. That was OK, Laurie said again; she just wanted to thank him for his role in contributing to her existence. She told him she wanted to meet him. He said he'd be willing, but he needed a little time to process this news and tell his wife and daughter.

For two days, Laurie was walking on air. Then she logged on and discovered the man's test had been deleted.

Laurie sent the man another letter. She included a screenshot proving their genetic relationship, in case he hadn't seen it. She was not trying to push her way into his family, she wrote, nor to make a claim on his money, and she was willing to sign a legal document to that effect. "I certainly hope you harbor no feelings of guilt or remorse," she wrote. "Everything turned out in both our lives exactly as God planned." What she sought were stories about her biological father and his family, to help her build a sense of where she came from. She would fly to him; she would treat him to lunch. One meeting, a few hours, was all she asked—which, depending on your perspective, might not be asking for much, or might be asking for a great deal indeed.

I first spoke with Laurie a few days after she sent that second letter, when it was becoming clear that her biological father was not likely to write her back. She told me all the things she'd ask the man if he'd talk to her: What books did he like to read? What had his childhood been like? Had he played sports back then? Had he been close to his parents? Did he have photographs of them so that she could trace her own face and those of her children in theirs? Could he tell her stories about his family? Because his family was hers now, and she longed to know her own family, if only a little bit.

She did not feel embarrassed about how she'd come into the world, she told me, and she did not regret her own discovery. She had been "devastated that I wasn't who I thought, that I was made of a stranger," but she was nevertheless grateful to know the truth. Still, she found herself imagining how this man, this stranger in all but genetic material, must be feeling. "We all tell ourselves stories about who we are, and we pick and choose what fits into that story," she told me. "I feel like I don't fit the narrative that he's chosen."

Who decides what story we get to tell? Countless American men have by now been contacted and told they helped make a child once, and that those children would now like to introduce themselves. Some knew this day might come; others never imagined it. Many more will face this reckoning in years to come. And then there are men who find out the children they raised are not their own. "Sometimes it's hard, 'cause she reminds me of what my ex did," one man told *The Atlantic*, referring to the girl he now knows is not his genetic daughter. The revelation led to his divorce. Genetic counselor Brianne Kirkpatrick, who has started her own secret Facebook groups for people who've experienced DNA surprises, had a client who discovered at the age of seventy-eight that neither of his children were genetically his. "He was *pissed*," Kirkpatrick says. He wondered if he could sue his ex-wife.

But just as you can't anticipate what you'll discover when you test, you can't predict how the person on the other side of that test will react. I emailed with a man named Jeff Lester in Lebanon, Missouri, who told me the discovery of a daughter he'd unknowingly conceived at the age of sixteen was a miracle. She'd matched him on AncestryDNA during a period when he'd put his genealogical work on hold, so he never saw her messages there; instead, she eventually messaged him on Facebook. Jeff told me he thought at first this was a scam. He was fifty years old, and he did not remember getting any girlfriends pregnant back in high school. But he logged onto Ancestry, and there she was, this girl—no, this woman, already thirty-two years old, already a mother herself—listed among his matches as his daughter.

Jeff told me he could not get over the unlikeliness of his daughter finding him. Outside of DNA testing, it almost certainly would not have happened. She had been given up for adoption when she was a baby and did not know the identity of either of her biological parents. When Jeff helped her uncover her mother's identity, she turned out to be someone with whom he'd had a brief relationship he didn't remember consummating. It "sounds like a bad after-school special," he wrote me.

The other piece of the miracle, Jeff said, was that he was not supposed to be alive. He'd been diagnosed in his twenties with ALS, also known as Lou Gehrig's disease, and had been on a ventilator for two decades. With this sword of Damocles hanging over him, he'd built a life for himself, marrying and having three wonderful daughters. He was astonished that he'd not only outlived the odds but lived long enough to discover that, in fact, he had *four* daughters, plus a son-in-law and grandchildren; he felt the hand of God at work. He told me all this over email and Facebook Messenger, typing by way of a wireless head-controlled mouse attached to his glasses, because the ALS severely limited his ability to speak.

Jeff told me he was determined to live as fully as he could while he was still on this earth. He had forged a relationship with his oldest daughter, and he and his wife even got to host the grandchildren for a sleepover. He did not feel guilty or ashamed about her existence;

those were wasteful emotions, and his disease had taught him that none of us have time to waste. Under the shadow of imminent mortality, "you focus more on what is truly important in life," he wrote. For Jeff, the unexpected outcome of recreational DNA testing simply meant more love.

6

ALICE AND THE DOUBLE HELIX

When Alice sent her saliva to 23andMe, she did not know exactly what would happen to the vial inside its cardboard box. It was like magic: The box went off in the mail, and the next time she'd know any more would be when she received an email telling her that results were ready. By now, the DNA of tens of millions of people has gone on this kind of journey, revealing its secrets to us. On a hot summer Thursday in Houston, Bennett Greenspan showed me what that journey looks like.

He took me down a flight of steps from his office. He showed off the sign for the women's bathroom, indicated by an XX, and the men's, a big X and a small Y. We walked past staff sitting at desks, and Greenspan handed me a white lab coat to put on and blue booties to stretch over my sandals. Then we went through a secure door and into a sprawling series of rooms filled with freezers and robotic equipment, the linoleum floor abruptly changing color to indicate where the space had been expanded as the company grew.

The air was filled with the whoosh and whirring of machines. FamilyTreeDNA analyzes the cells customers gently scrape from the insides of their cheeks with swabs that look like long Q-tips, and near the door, an auto-sorter with a robotic arm was reading the label attached to each vial of cheek cells in preserving liquid and arranging it according to the type of test the customer had ordered—whether autosomal, mitochondrial, or Y chromosome testing. Another set of machines transferred the liquid from each tube onto a plate designed for extracting DNA by means of a chemical process that "rips the cellular wall apart," Greenspan said. "Now we can get to the DNA in the center." In another spot, about twenty machines performed an enzymatic process called a polymerase chain reaction, allowing the lab to "make more needles than haystack" by "amplifying" the DNA to generate many, many more copies so that it can be analyzed.

We walked through to another room, where Greenspan showed me what looked like a small rectangular metallic bookmark, lined with two rows of smaller rectangles. The DNA of twenty-four customers could be placed on this specialized Illumina chip, one on each little rectangle, binding to it through a process called hybridization.

"There's seven hundred thousand data points on each of those rect-angles," Greenspan said—seven hundred thousand data points that could say A, C, G, or T. Nearby, along a row of metal shelving, eight laser scanning machines made by Illumina were reading these chips. In a minute and fifteen seconds, a machine could read all seven hundred thousand data points on a single sample; for an entire bookmark with its twenty-four samples, it took about a half hour.

Greenspan showed me a new robotic storage machine that had taken months to install and had the capacity for about two million samples of extracted DNA, ready for any additional testing custom-ers requested; an older freezer held another half-million tubes of extracted DNA. The company stores extracted DNA for several years and, in another place, keeps a second set of vials with their DNA not yet extracted; these do not degrade as quickly. "I've got millions of them downstairs," he said. He said the lab had the capacity to auto-sort, prep, and extract DNA from about ten thousand samples a day. I asked if he was ever operating at capacity.

Oh yes, he said. During Christmas.

* * *

What happens to your DNA after it gets extracted and amplified, hybrid-ized and scanned? There are several ways to answer that question. For one thing, it is analyzed; each company has its own method for deter-mining consumers' ancestral origins, which is why your results will vary somewhat across brands. We'll come back to that, because it's a big topic, wrapped up in questions of race, culture, science, and identity, not to mention how the legacy of eugenics is woven into our understanding of biological difference. Another thing that happens after DNA is processed in a lab is that these results are delivered to customers, who layer their own interpretations atop the proprietary analysis that companies like FamilyTreeDNA and 23andMe have performed. But Alice was still a few weeks away from being able to see her results, and from determining whether they jibed with her AncestryDNA test, or whether this whole foray into consumer genomics would turn out to be a waste of time.

So while we wait, it's worthwhile to consider a third way in which we might answer the question of what happens to your DNA. The answer, depending on where you test, may be that it becomes profitable in a very particular sense. And not just when you buy a kit for Christmas, but for a very long time after that.

The way that the value of those letters, those As, Cs, Gs, and Ts, increases once they enter the corporate arena, and the places those letters go—whether or not you realize it—has led some observers to wonder whether consumers really understand what they're giving up. At a fascinating conference on ancestral DNA testing in Toronto in 2018, organizer Julia Creet, a professor of English at Canada's York University, suggested that "the information that you're giving them"—the DNA testing companies, that is—"is much more valuable, I would argue, than the information that you are getting."

To understand the potential here, consider the example of 23andMe. One of FamilyTreeDNA's big promises to consumers is that the company will never sell genetic data to third parties. But 23andMe has a totally different model, and it helps to rewind to how that company came to be.

By the mid-2000s, a number of health-related genetic testing companies had popped up, offering tests for conditions like ovarian cancer and cystic fibrosis directly to consumers. As one molecular biology journal noted at the time, the United States was especially ripe for this "do-it-yourself diagnosis": Americans tend to embrace an ethos of individualism, and besides, they were already accustomed to "marketed medicine and paying out of pocket for healthcare services." When a former health-care analyst named Anne Wojcicki co-founded 23andMe in 2006, in the midst of this growing field, the company had two goals: to offer consumers access to their own genetic information so that they could better manage their health, and to create a vast database of genomic information for research. "Solve health," Wojcicki phrased its mission. 23andMe collaborates with outside researchers to study the data, and sells it to pharmaceutical companies. In some cases, it gives away spit kits for free to people who suffer from diseases it wants to study. The company has said that this huge repository of genetic data, along with information participating consumers have provided, allows the company to move much faster and more cheaply than such research usually requires.

It goes without saying that genetic health information can be incredibly empowering. When the actress Angelina Jolie revealed in 2013 that she'd undergone a preventative double mastectomy after discovering she was a carrier of the *BRCA1* breast cancer variant, the news caused consumers to rush to 23andMe to buy their own kits. But 23andMe's business model, which it has described as a "virtuous circle," has also been subject to criticism and suspicion. The year after Jolie's op-ed, *New York* magazine described the company's approach of charging people for access to their own genetic information as "a sleight of hand so quintessentially American that Tom Sawyer might have dreamed it up."

In 2017, after a big round of funding, *TechCrunch* estimated 23and-Me's value at $1.75 billion. "The long game here is not to make money selling kits, although the kits are essential to get the base level data," Patrick Chung, a board member and venture capitalist, told *Fast Company* in 2013. "Once you have the data, [the company] does actually become the Google of personalized health care."

The vast majority of 23andMe's customers consent to have their de-identified data used for genetic research, allowing the company to boast that "23andMe's research platform is currently the world's largest consented, re-contactable database for genetic research." Customers can consent to different degrees of having their information shared, whether in aggregated or in individual form. They also fill out surveys, answering questions about topics as varied as arthritis, skin color, body mass index, psoriasis, cancer, multiple sclerosis, rosacea, and cholesterol. The addition of this information provided by the subjects of the genetic research themselves gives that data much-needed context for researchers.

"We specialize in capturing phenotypic data on people longitudinally—on average three hundred data points on each customer," Wojcicki reportedly told a room full of researchers at a company event in 2018. "That's the most valuable by far."

The company's consent document explains that it may develop intellectual property or commercialize products and services based on studies it conducts and "in such cases, you will not receive any compensation." But it frames the invitation to participate as "becoming part of something bigger," and indeed, for consumers, the impulse to share often stems from magnanimity: If something in my genetic material could help someone else, have at it. With all that crowd-sourced genetic data, Wojcicki has said, "I can partner with academic researchers, I can partner with biotech companies, pharma companies, to say how can we all—who have a common interest—come together and say, I want a solution?"

Along with research and funding partners including Genentech, Pfizer, NIH, Stanford, the National Institute on Aging, and the Michael J. Fox Foundation, the company has collected data and produced research on, among other things, Parkinson's disease (Wojcicki's former husband, Google co-founder Sergey Brin, found out he was a carrier after taking a 23andMe test), allergies, nearsightedness, breast cancer, motion sickness, rosacea, migraines, depression, inflammatory bowel disease, and health disparities among African Americans. In 2015, the company launched 23andMe Therapeutics, a research and drug development group. By 2016, *MIT Technology Review* reported, the company had already sold access to its vast treasure trove of consumers' genetic information to more than thirteen drug companies.

Ancestry has had research partnerships, too. For a time, it was collaborating with a Google-founded longevity research company, Calico, to "research the genetics of the human lifespan." It has also collaborated with the nonprofit University of Utah. Kristen V. Brown, the Bloomberg journalist, said it made sense to her that a company like Ancestry would be looking beyond genetic genealogy. "If you want to expand your market, you eventually want to expand your product offering," she said. "I think the health market is much bigger." Brown was right. The space was heating up. In 2019, both Ancestry and MyHeritage announced they were moving into the arena, teaming up with physicians to oversee their new health tests, which meant the companies didn't need to get FDA approval to sell directly to consumers the way that 23andMe has had to.

But there are unresolved tensions about fairness, about money, and about how much everyday consumers of these services understand. When 23andMe announced in 2018 that it had entered into a collaboration to give British pharmaceutical behemoth GlaxoSmithKline access to its customers' genetic data as part of a four-year deal to (in Wojcicki's words) "accelerate the development of breakthroughs," with Glaxo-SmithKline purchasing a $300 million stake in the company, ethicists lined up to voice concerns. Was that genetic data still safe as it was changing hands? Did people who signed up to participate in research truly comprehend how asymmetrical their relationship with 23andMe was—how much it and other companies stood to profit off the information they and millions of others had paid for the privilege of providing?

"Making money off our genetic data?" one outraged consumer of many tweeted in response to the GlaxoSmithKline deal, adding that he was revoking permission for 23andMe to use his data for research. "This smells like a Henrietta Lacks situation," someone else replied.

* * *

What is consent?

As one lawyer and bioethics researcher pointed out in the clamor following the GlaxoSmithKline deal, 23andMe did appear to be disclosing how it could use consumer data in its consent documents—"the challenge is that people don't read it." Indeed, all the major DNA testing companies state in their policies that they don't sell or share genetic data for research purposes without customers' explicit consent. Companies with research programs generally make them optional. You can also change your mind about participating, though your information can't be removed from research already under way. 23andMe allows that, in addition to conducting research with consumers' consent, it does "use and share aggregate information with third parties" but says it goes to great lengths to ensure that it protects consumers' private information.

But the problem is, it can be difficult for consumers to understand exactly what they're signing up for—both how their data could be used and how it is being protected. I spent many hours poring over the companies' language about privacy, and then diving deeper into their various policies, all of them thousands of words long, trying to parse the definition of terms like "personal information," and thinking how much easier this would be if I'd gotten that law degree. Do such contracts make sense given the way we behave online? One 2008 study estimated that if American Internet users attempted to read the privacy policies they regularly encounter online word for word, they'd spend upward of two hundred hours a year doing so.

"We have all been conditioned and socialized into 'I agree,' right? Get me onto the next page: 'I agree,'" philosopher and bioethicist Françoise Baylis has said. And it's true. A few years back, professors conducting an experiment gave 543 college students the opportunity to sign up for a fictitious social networking service called NameDrop. The students largely bypassed the site's lengthy Terms of Service and Privacy Policy, missing the part where they were agreeing to hand over their firstborn children. Poor regulation and overly long and complex contracts become even more of a problem with sensitive genetic information. In a 2018 study looking at the policies of ninety American consumer genetics testing companies, Vanderbilt University privacy researcher James Hazel found varying and often vague information about how long companies would retain genetic information, who owned it, and whether and how it would be shared with third parties. In the event of a sale, merger, or bankruptcy, the information might "be treated as an asset that would be transferred to the acquiring entity." And the companies' language also typically indicated they were permitted to change their policies whenever they liked. Privacy issues within the industry are largely self-regulated. Most direct-to-consumer genetic testing companies are not governed by the Health Insurance Portability and Accountability Act, or HIPAA, which protects the privacy of genetic data in clinical and research environments. Instead, they are regulated "by a patchwork of federal and state laws," writes Hazel.

Even Paul B. Allen, one of the founders of Ancestry back when it was just a genealogical subscription service, told me that while he thought his old company and 23andMe were developing amazing features with DNA testing, "I'm worried about the control of the data, and I'm worried about corporations owning this data."

And what if the data at one of these companies—or at a third party with whom they share the data—were to be breached? In 2018, this happened to MyHeritage. The email addresses and encrypted passwords of ninety-two million users (though, notably, not their DNA data, family trees, or credit card information) were found sitting on a private server. There have been breaches of other big data companies, including Facebook, Equifax, and Google. To be fair, just how informative that data would be is unclear. Debbie Kennett, the genetic genealogist and researcher, told me she thinks the information people share on Facebook and through things like supermarket customer loyalty cards is more revealing.

"If I was to send you my raw DNA data sequence, on its own, it's not going to tell you a great deal," Kennett says, even if the data were uploaded to a third-party health site like Promethease or to a genealogical research site like GEDmatch. "I think people expect the DNA to tell them more than it actually can, and they expect it to be predictive of health issues, which it is not," she says.

Still, there are larger, almost existential questions about how we treat information that clearly has great value once it is parsed and put in context. It has the power to tell us who are relatives are and aren't, and to reveal what diseases we may be at elevated risk of, and—here in the United States, at least in theory—to make it difficult for us to obtain certain kinds of insurance. (I'll come back to that.) Some privacy scholars believe HIPAA's protections should be extended to cover companies like Ancestry and 23andMe. Bioethicist Thomas May wouldn't go that far, but he thinks more could be done to protect people's information. The direct-to-consumer world is operating like the wild west, he's written: "In the health care context, we would not be so lax about a sample's chain of custody, nor return test results without following clear procedures for establishing who has a right to such information."

In part, this is the cost of the empowerment that Anne Wojcicki envisioned. Consumers themselves are often handling one another's genetic information; if they submit a spit test to 23andMe, for instance, they are stipulating under its terms of service that they have "legal authority" to do so. The major companies make it possible for parents to consent on behalf of their children—a fact that profoundly troubles some bioethicists for reasons having to do with informed consent, with children's autonomy, and with the potential for years of anguish and anxiety on the part of parents *and* children should they learn something troubling. One blogger wrote about devising a method perfect for infants, the elderly, and anyone who couldn't or wouldn't generate the needed liquid for companies like AncestryDNA that require saliva samples. She made a video demonstrating her method for mixing a cheek swab with faked saliva on her infant nephew. Commenters beneath this thanked her, claiming her method had worked for them.

Elsewhere online, prominent genetic genealogists explain to seekers how they can swab their dead family members and send the samples to FamilyTreeDNA—and one blog chronicles how, in an instance when the cheek swabs failed because too much time had passed since death, hair follicle samples worked. In an attempt to create some guidelines, leading genetic genealogists have written a set of standards for their community, stipulating that testing DNA requires consent. And meanwhile, some companies have announced that they're offering, or plan to offer, what's known as "artifact testing" for old envelope flaps and stamps, perhaps from love letters sent between grandparents, raising profound questions about the ethics of testing those who can't consent because they're no longer alive. MyHeritage founder Gilad Japhet, who collects autographs, has spoken of planning to test the DNA on envelopes he owns sent by dead notables. "Their DNA is coming to MyHeritage very, very soon," Japhet announced at a conference in late 2018, "and I think it will be amazing for people to find if they're related to Winston Churchill or to Albert Einstein."

The wild west, indeed.

* * *

One day late in the summer of 2012, Alice got an email from 23andMe. Her results were back. She went to the website and logged in. She navigated to her list of DNA relatives. Her strongest connection, with about 12.5 percent shared genetic material, was to a man she hadn't known she'd find in the database: the son of her brother Bill.

Alice realized in an instant what this meant: Her nephew was her full genetic nephew and her brother was her full genetic brother. She was not adopted. She was not the product of an affair. As far-fetched as those ideas had seemed, she could not be certain till this moment. This meant Jim Collins, the kind, taciturn, hardworking man she'd loved, was indeed her biological father. She felt tremendous relief.

In retrospect, of course, it seemed impossible that Alice Collins Plebuch was *not* the daughter of Jim Collins. Not only does she share his eyes, his cheekbones, his wide feet, his stature, and his smile, but she and much of the Collins family inherited her father's facility for numbers, for systems, for keeping track of vast quantities of stuff. "My dad was pretty much a mathematical whiz," Jim's son Jim the younger told me. "When I was in college he had only gotten his high school equivalency, but he was able to do my calculus." Jim seemed to understand math on an intuitive level, though he couldn't necessarily explain his methods if you asked him to. He did it all in his head. Sometimes, for fun, his kids tested him. "You could have a row of numbers this wide and this long," Alice told me, making a square with her hands about the size of a dinner plate, and "he could add them up and get the result faster than you could read them or put them into a calculator. He would always be done before you, and he would be right."

This talent has trickled down: Jim the younger is a retired software and systems engineer who at one point worked on NASA supercomputers. Another brother, John, majored in electrical engineering and then worked in systems design at IBM. Brian was a materials manager who did purchasing, warehousing, and shipping. Bill was a facilities supervisor at a large research center. Ed managed warehouses, analyzing and overseeing the computer ordering and fulfillment systems. "It's some form of synthesis," Alice's best friend, Kathy Long, told me. "They take a lot of data and they synthesize it down to bite size, whereas the rest of us don't quite organize things in the same way." And then of course there's Alice, who during lunch one day with me and her niece, Cherie Collins, explained how, for purposes of genetic genealogy problem-solving, she used to keep a bar chart of a certain set of variables in her head. "You can see it's passed on—whatever he had," observed Cherie, who owns her own web development company and appears to have inherited Jim's particular mental prowess, too. (Gerry, a teacher-turned-homemaker, once joked that if any of the Collins kids was going to turn out to *not* be a Collins, it would be her; she was the only one of them "not on the cutting edge of technology.")

Jim the elder excelled at these kinds of things during his twenty years in the army, first at West Point and then all over the world: learning and teaching military engineering, and overseeing a system to purchase, store, inspect, repair, pack, and issue supplies, tracking it all on thousands upon thousands of stock record cards. The maintenance of all this—of records, of workflow—was not unlike what Alice did in her career with benefits, only Jim didn't have the opportunity to automate things using computers and the Internet. Outside the army, he was incredible at card games because he could keep track of everything that had been played and seemed to see through the cards in the other person's hands, a tactic that made him a rather unpopular opponent. Sometimes he had to lose on purpose so that people would continue to play with him. He also played poker and bridge and dice, and killed it at pool. After the orphanage and before the army, during the "misspent youth" he sometimes liked to talk about, he ran numbers for bookies managing illegal gambling rings, collecting money and helping keep their records.

But back in 2012, the results from 23andMe confirming that Jim Collins really was her father were nonetheless a big deal to Alice. Reassured by her connection to her nephew, she looked to see what the company could tell her about her ancestral background. Would it tell her AncestryDNA's ethnicity estimate was wrong? But instead of clarity, she found the results hard to parse—far more difficult in those early days of the industry than they are now. As best as she could tell, from staring at an image the website showed of her twenty-three chromosomes, with little flags showing the apparent geographic areas her genes hailed from, she had a lot of Northern European ancestry, as well as ethnicity from places in Europe including Lithuania, Belarus, Ukraine, and Russia. She could not tell just how much Eastern European ancestry might have been passed down to her—or how far back in time she'd need to look to solve her mystery.

She emailed her nephew, who informed her that his dad, Alice's brother Bill, had also tested through 23andMe; he just wasn't showing up as a relative because he hadn't turned on a feature to make himself findable. Bill Collins had tested, it turned out, at his wife's behest, but hadn't been interested in his ethnicity results, figuring they were junk science.

Alice sent an email to the professor in Illinois whose name she'd seen on the Yahoo group DNA-NEWBIE. "I'm so confused," she wrote. "For the last sixty-four years I've been 75 percent Irish and the remaining 25 percent split equally between British and Scottish." AncestryDNA had told her she was something else, and now 23andMe's results were proving difficult to interpret. "Would you please run it through your analysis?" she wrote. "My genome is attached." She had accessed her raw genetic data from 23andMe, downloaded it, and attached a zipped file, about 8.2 megabytes containing hundreds of thousands of As, Cs, Gs and Ts, possible clues to an answer if someone in possession of the right tools and scientific understanding could make sense of them.

She also downloaded her raw data file and uploaded it to GEDmatch, the free, publicly accessible genealogy database, along with her brother Bill's 23andMe results. Several years later, GEDmatch would become famous—infamous, some might say—when members of law enforcement and genetic genealogists began searching the genetic data of the million or so people on the site to solve particularly heinous cold crimes. But back then, it was a little-known destination for early adopters of this nerdy hobby, offering additional tools for people to analyze their genetic data, and a place for people who'd tested at different companies to compare their results.

These days, newcomers to their own DNA can fairly quickly get up to speed on how to understand their own ancestral origins and their relationships to genetic relatives. There are ethnicity pie charts with increasingly precise estimates of biogeographical origins. New tools at AncestryDNA and elsewhere make it easier to understand exactly how two people are related. The many Facebook groups geared to this interest are stocked with fellow citizen-scientists ready to share their tips. But in 2012, the knowledge was harder to gain and the process much slower. Alice had to understand the science behind all of this, so she studied up on SNPs, on the distinction between two people sharing genetic segments because they're related or by chance, on how to understand relatedness, and on the number of centimorgans—a centimorgan, or cM, is a unit for measuring genetic linkage—she could expect to share with certain relatives. As geneticist Ricki Lewis explains, genetic linkage reflects the way inheritance works, with DNA sequences that are near each other on the same chromosome tending to stay together as they're passed down. These likenesses are translated into patterns that help geneticists see evidence of biological relationships.

And Alice began to connect with her DNA relatives, asking to see more of their genetic material in hopes she could understand her own.

One DNA relative at GEDmatch, a man living in Canada, messaged her. They spoke by phone, and the man gave her an impromptu seminar, of sorts; he was the first of many genetic relatives moved by her mystery and eager to help. He walked her through how to compare the overlapping genetic data along her chromosomes with relatives at 23andMe, and he recommended she try testing a brother's Y-DNA at FamilyTreeDNA. Plus, he told Alice that, if the ethnic mix AncestryDNA was giving her was correct, she was likely descended from Ashkenazi Jews, a term for a group of Jewish people who trace back to central and eastern Europe.

Alice asked her Canadian cousin a question: *What was that word?* She'd never heard the term "Ashkenazi" before. It took her months to figure out how to pronounce it properly.

Judaism is, of course, not just a religion; it is also a culture and an ethnicity with a distinctive genetic signature, which explains why Alice could be considered ethnically Jewish even if she was not practicing the faith. "The Ashkenazi gene pool looks half European and half Middle Eastern, but it's unique in its combination thereof," says University of Arizona's Michael Hammer. Population genetics papers by Hammer and others trace the geographic journey of this community and document how its unique history, including a long-ago population "bottleneck," gave Ashkenazi Jews an unusually high prevalence of certain disease-associated genetic mutations. As Ashkenazi Jewish people traveled, and as many were chased eastward in Europe by persecution, they married among themselves, remaining "distinct from the local community," Hammer says. They suffered mass killings in the crusades, in pogroms, and, of course, in the Holocaust.

After several hours of genetic briefing by the man from Canada, Alice hung up the phone. In coming months, the enormity of the task in front of her would become clear. There were many, many genetic relatives like this man, and each one of them might be a clue. At heart, she had a data problem. First, she had to figure out exactly what her ancestry was, and—if she were indeed of Ashkenazi Jewish descent, as it appeared—which parent was passing down this history. And then, she might have thousands of genetic relatives, linked to her by count-less segments of shared material. She had to figure out how to organize it all, how to separate the useful information from the dross, how to compare the useful bits, and how to use all those bits to figure out, if she could, where her people came from and how her family had gotten its history so wrong.

Thank goodness for that analytical brain—Jim's legacy. She was prepared for this. During her career at the University of California, she had excelled at organizing the flow of information. In the late 1990s and early 2000s, when she was in charge of "employee direct systems," she was tasked with overseeing the interactive voice response system that people had to use to enroll for benefits (remember those?), with its pitiless choose-your-own-adventure logic: press 1 for this option, press 2 for that. She hated it, and fought to move enrollments to the web—and then, once everything was online, what a pleasure *that* was. By god, she had wrestled with the phone tree and won! She could do this. She loved numbers, and statistics, and logic, and she was retired. She was more than ready for this moment.

Alice Collins Plebuch was ready to build a database of her own DNA.

7

EUREKA IN THE CHROMOSOMES

Linda Minten's childhood was lonely. She describes her mother as self-involved and cold, her father as loving but beset by poor health and a painkiller addiction. Her parents split up when she was six. Linda lives now on a farm in rural northwestern Oregon, where she has a family, raises and sells a few beef cattle, grows timber nearby, and acts as a kind of keeper of lost creatures not unlike the child she once was. They include a blind chicken she takes for daily walks, and a goat she bottle-fed and raised in her house.

In her early fifties, Linda began experiencing symptoms of what her doctor believed might be ovarian cancer. She was bloated, bleeding, and in pain. An ultrasound was inconclusive, but testing found an elevated level of a tumor biomarker. The fact that Linda had recently learned she was likely about half Ashkenazi Jewish through her mother worried her, because it increased her likelihood of having a *BRCA1* or *BRCA2* variant, which would put her at higher risk of that kind of cancer. Wondering if her father might also have Ashkenazi Jewish ancestry, further increasing her risk of these variants, Linda decided to find out more about her ancestral heritage through AncestryDNA. She also ordered a clinical-grade test from a company called Color, to look for many mutations that increase cancer risk. Linda's father had passed away decades earlier, but Linda told her mother what was going on and consulted her on the family's health history, which included two maternal relatives with cancer and the fact that Linda's mom had had a hysterectomy. Linda brought all this information back to her doctor, who considered all the risk factors and recommended she have an emergency radical hysterectomy. "We couldn't wait for it to pan out," Linda told me.

The Color test came back the day before Linda's scheduled surgery. It didn't find evidence for the genetic mutations Linda was worried about, but her doctor felt the procedure was still warranted. So she had the surgery, and afterward, her doctor told her there was no evidence of cancer in the organs that had been removed. Linda was relieved and felt they'd made the right decision based on the information they had. But a week and a half after the surgery, as she was recovering, her AncestryDNA results came back, and that's when things took a turn. Her relative matches listed a man she didn't know as her father, as well as a half-sister and cousins she'd never heard of.

Understanding didn't come right away. Linda had to piece the clues together. She had to call her mom, who said she didn't know why the test had predicted some stranger was her father. She had to call Ancestry and speak with a kindly customer service representative, who assured Linda that the results were almost certainly accurate and that she was far from the first person to call with unexpected results. She had to log back onto Ancestry.com and look at her ethnicity, which suggested she had no Jewish ancestry at all. She had to examine the language on her birth certificate until she realized it had been amended and issued a year after her birth. She had to call the county in Arizona where she'd been born, and find out that her original birth certificate was sealed. And, she says, she had to call one of her brothers and wring the truth out of him.

This was how Linda found out she was adopted—a plain fact she says everyone, including neighbors, teachers, and her two older brothers, had known, but which she herself had never been told. Linda was fifty-one years old.

And she was dumbfounded. Dumbfounded that her mother had lied when asked, and hurt that her brothers hadn't intervened despite their mother's edict not to. "I just went through a hysterectomy," she told one of them. If she'd known the truth, she wouldn't have been so panicked by what she thought was an elevated risk. "I would have had more time. I could have retested." She might not have needed the surgery at all.

It was strange. Linda had never suspected she was adopted, even though as a child, she'd sometimes fantasized that she was. Now, the revelation of her adoption was freeing, making her mother's seeming dislike for her feel less personal. When she spoke with her mother again, Linda told her what she knew, and she says her mother told her, "We did the best we could." Mary Rubens, a close friend to Linda who is also the ex-wife of one of her older brothers, confirmed the basic timeline of Linda's story and told me that even *she* had known Linda was adopted since she'd been a teenager dating Linda's brother. Mary had long expected Linda's family to tell her, but she herself did not, feeling it was not her place, and feeling bound by a promise she'd made to Linda's father decades earlier that she never would. Once upon a time, some thought it best that adopted children not be told they were adopted. "I just think that he didn't ever want her to feel unloved," Mary told me.

And now what? At once distraught and relieved, and eager to understand where she'd come from, Linda set out to find her birth parents. She started with her birth father, who was already in the AncestryDNA database under his initials. She thought he might have tested his saliva in hopes of finding her. But the first few times they spoke on the phone, the man's reaction surprised her. "'I'll be darned,'" she says he remarked. He said he hadn't known Linda existed, nor did he remember her mother. Linda told him she was conceived in March 1965—did that ring a bell? Nope. Her mom might have been a redhead, like Linda—did that help? It did not. In their second conversation, he told Linda that these phone calls were getting him in trouble with his wife, and she apologized, feeling again like the child who was unwanted and out of place.

Still, she kept trying for connection. "Look, it's been a month," she told her biological father. "If you're not going to tell your kids, I will." One of her father's daughters was already matching her as a close relative in the database, and on Facebook she'd become friends with several paternal relatives. She figured it was only a matter of time before her father's three daughters discovered her existence. "I knew what being the last person to find out a secret felt like," she said. So she sent them a note.

All I could think when I read the Facebook Messenger exchange between Linda and her half-sisters was what a jarring and thoroughly modern way this was to meet one's genetic family. Linda's tone is effusive, vulnerable, alternating between being eager to know her siblings and allowing they might want nothing to do with her. She explains that she was conceived before their parents were married, but she's aware that her existence may nonetheless cause distress to them and their mother. Their replies are terse, stunned. Later, Linda would say she was so excited to have half-sisters that she didn't fully appreciate how her story posed a fundamental threat to *their* story.

Linda also set out to figure out her biological mother's identity, working with family trees she created based on maternal cousins among her relative matches, and homing in on a woman who was a classmate of her biological father's. When she figured out the identity of the woman she believes was her biological mother, she learned the woman had passed away six months earlier, leaving Linda to piece together a biography from clues: an obituary, which she read countless times, a photograph on Classmates.com, and conversations with maternal relatives. "You don't get to know the person, so you hold on to these little tiny scraps," she told me. The woman also left behind two grieving daughters who said they'd never known of Linda's existence. As with Jacqui's experience, the news Linda bore appears to have contributed to poisoning her overtures. She forwarded me an email sent by the husband of the deceased woman, calling Linda "deplorable."

Linda kept pursuing various members of both families, certain that if she could just reach the right person, she would be heard and, perhaps, welcomed, but for the most part, she's been shut out. She says she's heard that some in her father's family believe she is interested in money, an accusation that deeply wounded her. She forwarded me a "discontinuance of contact" letter sent by an attorney for her biological father and his wife, telling her that his clients believed her "assertions are based on unproven guesses and conjecture," and asking her to "slow down and research this in a more measured, analytical manner."

Negotiations over having Linda and her father submit to formal paternity testing broke down amid questions of who would pay for what, and mistrust on both sides.

"When you're fifty-one years old and this hits you, you feel like you're six," Linda told me one day, her voice breaking. "You just feel so vulnerable, so raw and vulnerable." She had a neediness I'd come to recognize from other seekers who felt they'd been denied their identities; entire conversations were conducted in a minor key. Linda told me she understood why her mother had given her up for adoption—"she was young, scared, and the love of her life just married someone else." But being kept hidden hurt so bad. Perhaps the way she'd come into the world had once been taboo; perhaps her mother's pregnancy had been a problem to be taken care of. But she herself did not want to be a taboo or a problem. She did not want to have to apologize for her own existence.

* * *

In this moment, we are learning just how painful and damaging it can be to discover that the truth of your genetic identity has been kept from you. One study out of the United Kingdom, looking closely at the experiences of sixty-five people conceived by sperm donation, found that nearly half hadn't been told how they were conceived till adulthood. For some, this created relationship problems with their families, and especially with their mothers, who they saw as having had a particular moral responsibility to tell the truth. They were, as the study put it in dry academese, "frequently more concerned about prior parental deception than by their parents' use of donor conception." However well intentioned their parents' decision was, these subjects experienced the information kept from them as a particularly "toxic" kind of omission.

Psychologist David Brodzinsky told me that in cases he's encountered over the years of patients who come in grappling with not having known until adulthood that they were adopted, "almost universally, it's the lie" that troubles them, rather than the fact of having been adopted.

"Relationships are built on trust. If you go through life thinking you are who you are, that your family is your family—that is the rock bottom foundation of who you are—and you suddenly realize, 'Wait, I have a different family, at least biologically; I've been lied to all these years,' it can be extraordinarily upsetting," he says.

There's a profound cognitive dissonance for the seeker whose revelation comes abruptly, accidentally, and well into adulthood. And on the other side of the secret there is sometimes a whole architecture built around the denial of the plain biological fact of someone's existence. Big secrets can be corrosive, and in the wake of their discovery, families can come apart.

"It's a secret, but at some point it becomes a lie, too," Linda Minten told me one of the first times we spoke. When I touched base with her again almost a year after our first conversations, she told me she hadn't spoken with her adoptive mother in well over a year.

* * *

Genetic revelations can be life-changing. This is what the early adopters knew before the rest of us had even heard of 23andMe. If questions of ethnicity and family relationships can be dramatic, revealing whole new categories of relationships—genetic siblings who deny their siblingship, for instance—health-related results are at least as profound. It's just that they're often also harder to understand.

Genetic counselors will repeat this till they're blue in the face: The results of health-related genetic testing can be very difficult to interpret. Bioethicist Thomas H. Murray points out that when people think of genetic risk for disease, they often conjure up examples like Huntington's disease, a rare inherited disorder in which the brain's nerve cells degenerate, typically causing uncontrolled movements, difficulty thinking, and psychiatric problems. Having a mutation in just one of the two copies of the relevant gene is enough to cause the disease. A person with the mutation will almost always develop the disorder.

But the vast majority of diseases in which scientists believe genetics plays a role are complex disorders, in which many mutations (or variants) in combination with each other and with the environment appear to play a role. The individual contributions of these variants tend to be subtle on their own. In other words, in most cases, health-related genetic testing is probabilistic, rather than deterministic; it can't tell you for certain that you have or are going to develop a disease. A positive result for particular mutations may simply indicate an increased risk of developing something—and how meaningful that increased risk is will vary. A negative result may not mean there's no risk for that disease. Other variants, as well as factors like a person's medical and family history, what they eat, whether they smoke, and what medications they take, will weigh into determinations of risk, which is why many doctors say health-related genetic testing should take place in the context of a medical consultation. It can be difficult to understand this kind of risk. "We have a hard time understanding probabilities—it's not at all an intuitive way of thinking," writes cultural psychologist Steven J. Heine.

23andMe's testing illustrates the complexities of understanding genetic risk. Beyond the three *BRCA1* and *BRCA2* gene variants that the company tests for, there are about a thousand other potential variants in those genes, any one of which could, on its own, elevate the risk of cancer. And beyond that, mutations in other genes can put people at elevated risk of hereditary breast cancer. " 'F.D.A.-approved' does not necessarily mean clinically useful," the *New York Times* editorial board wrote in a criticism of 23andMe's testing approach.

23andMe argues that more information is better, and says it has worked hard since its time in the FDA-imposed wilderness to make this information comprehensible, and to communicate to consumers the context and limitations of its testing. (For late-onset Alzheimer's and for Parkinson's disease—"serious conditions that do not currently have effective treatments or cures"—23andMe customers must explicitly opt in.) The company told me many customers who've tested positive for *BRCA1* and *BRCA2* gene variants would not have otherwise known they needed to worry; they were unaware of Ashkenazi Jewish heritage that put them at higher risk, or didn't know of close relatives with cancer. One customer, who was told by her doctor that she didn't need to test for these mutations, discovered through a 23andMe test that she had a *BRCA1* gene mutation that put her at high risk. She had a medically-recommended double mastectomy after confirming the finding with medical testing, and credits the company with saving her life.

"Our philosophy with genetics and science is anyone can be a scientist," 23andMe's Anne Wojcicki said in an episode of her company's podcast, *Spit*. "Here's your genome, you can learn about it." She has framed the scientific community as somewhat elitist, and has suggested those pushing back against the delivery of health care information directly to consumers are reminiscent of paternalistic fears in the 1960s and '70s that women couldn't handle at-home pregnancy tests. But scholars and medical professionals worry about the unintended consequences of consumer genetics. They warn about what one calls the "Walking Sick," who are inclined to skip screenings after consumer testing or forgo making lifestyle changes because of a false sense of security, and the "Worrying Well," who are driven to take more tests and perhaps, undergo unnecessary procedures. In a *New York Times* opinion piece, a genetic counselor wrote about the case of a young man who uploaded his raw data from 23andMe into Promethease, a third-party service that compares personal data against a database of medical literature. The disturbing results indicated a mutation associated with what the man understood to be a type of inevitable early-onset Alzheimer's. But after contacting a doctor, a geneticist, and a genetic counselor and finally getting a clinical test, he learned he didn't have the mutation after all.

23andMe told me that when it issues reports for the health risks that the FDA has given it permission for, like specific *BRCA1* and *BRCA2* gene mutations, it validates the results for accuracy in a number of ways. But the raw data are not validated in this way, and there can be errors. The company warns customers of the limitations of these "uninterpreted raw data," but it seems almost inevitable that more Worrying

Well will be created by an environment in which we're all supposed to be scientists. A small study found more of these errors, both within the raw data issued by consumer genetics companies and within the interpretation of those results by third-party services.

A number of health-related genetics testing companies have a hybrid model that requires physician involvement even as they market directly to consumers; some of them have begun to move toward working more with large health care institutions. Bloomberg's Kristen V. Brown believes that over time, there will be more convergence between the consumer genetics testing space and the medical establishment. "This kind of information about your health is never going to be not complicated," she says, "and you need an expert to help you interpret it."

* * *

There's another privacy issue inherent in genetic testing, given its power to reveal the risk of future disease. Both AncestryDNA and 23andMe say they don't share information with insurers or employers but warn that, in the event of a security breach, if genetic data could be linked to a person's identity, this could impact that person's ability to get certain kinds of insurance.

That's because in the United States, protections against genetic discrimination are imperfect. In 2008, Congress passed the Genetic Information Nondiscrimination Act, or GINA, a kind of preemptive civil rights law prohibiting discrimination based on genetic information in employment and health insurance. But there are loopholes in the federal law, as well as areas it doesn't even attempt to cover. For instance, GINA doesn't protect against discrimination in life, long-term care, or disability insurance, and, as *Fast Company* has reported, these types of insurers can ask "about health, family history of disease, or genetic information, and reject those that are deemed too risky." Only some states have laws that attempt to address the gaps in GINA.

"One of the situations I fear is, I apply for life insurance and they ask me, as they are wont to do, to sign HIPAA-compliant authorization releasing my medical records," says Mark Rothstein, a lawyer and bio-ethicist at the University of Louisville and author of *Genetic Secrets: Protecting Privacy and Confidentiality in the Genetic Era.* "Well, I haven't been tested by my doctor—but I've had a 23andMe test. And then they say, 'Have you undergone a genetic test in a research setting or direct-to-consumer?' And if I say no, I've committed fraud and they may deny paying me. And if I say yes, they say, 'Great, send it.'" Indeed, 23andMe warns of this very possibility, saying that failure to disclose health-related genetic information when asked by an insurer may be considered fraud. "I'm not aware of it happening, but it's just a matter of time," Rothstein says.

Advocates for genetic research worry that this threat may contribute to a chilling effect, discouraging people from participating in stud-ies and being proactive about their own health. And insurers fear that their model may collapse if lots of people who've undergone predictive genetic tests and know they're at high risk of developing certain dis-eases start taking out plans.

The idea of insurers using the possibility of future inherited dis-eases to deny us coverage is deeply discomfiting, dystopian even. Is genetic information sufficiently different from other kinds of medi-cal information—more precious, more private, more predictive of future health—as to warrant greater protections than other kinds of health-related information? This argument, known as "genetic excep-tionalism," is at the heart of GINA, and it's been the subject of long and heated debate. I sought perspective from Thomas Murray, the bioethi-cist who once wrestled with misattributed paternity in the case of the man with failing kidneys, and who many years ago chaired a Human Genome Project task force assigned to look at issues related to genetic information and insurance. Murray said that when he started on the task force, he was a proponent of genetic exceptionalism. The argu-ments in favor seemed compelling: Genetic information is a particu-larly personal and powerful kind of data. It is information that can serve as a kind of coded "diary" of a person's future health state; information that can reveal things not only about ourselves but about our relatives as well; information that, if it falls into the wrong hands, could give others the opportunity to discriminate against us. Insurers, sure, but maybe also totalitarian states.

And yet, as Murray turned the arguments over and over, he found himself unpersuaded that genetic information warranted special protection. After all, nongenetic information, like cholesterol levels and a penchant for skydiving, could also serve as risk factors in considering future health. And genes are not fate when it comes to disease; they are typically probabilistic. Murray worried that if he argued that genetic information was so different as to require its own class of protection, he'd be effectively arguing for a kind of genetic determinism, a mindset that genetics matters more than the countless other ways in which we define our own identities—an echo of what Nelkin and Lindee had warned about in *The DNA Mystique*. "Genetic information is special because we are inclined to treat it as mysterious, as having exceptional potency or significance," Murray concluded. He told me he was not opposed to GINA protecting genetic data, but that it didn't go far enough: The government should offer better protection for all medical information, not just the sort related to DNA.

The question of genetic exceptionalism is relevant to realms beyond the medical. Leading genetic genealogists argue against the idea that genetic information is so much more revealing and disruptive than what can be learned by traditional genealogy that it deserves to be treated differently. After all, CeCe Moore told me, genealogists could and did uncover family secrets back in the days when they worked with paper records alone. Looking at marriage and birth dates, they'd long noticed a number of children born after only seven months' gestation—implying either a surprising number of preemies or, more likely, a whole lot of shotgun weddings. And, genetic genealogist Blaine Bettinger has argued, when a state passes a law allowing adoptees access to their original birth certificates, it proves the point that "DNA does not have a monopoly on revealing family secrets." Besides, family members in the know have always had the ability to reveal secrets to one another, a fact of which adoptee Linda Minten is only too aware.

Still, as a practical matter, the amount of effort involved in making many such discoveries through old-school methods meant that in the pre-genomic era, genealogical revelations were limited to a smaller number of people—serious hobbyists who presumably became familiar with the risks inherent in their hobby as the years crawled by. You had to really want the truth back in the old days. Now the bar to entry is low, and many of the people testing are not familiar with the implications. The same factors that make autosomal DNA testing so powerful for helping researchers break through brick walls in their family trees also mean DNA can be much more jarring to people who otherwise would never have questioned their place in those trees.

Besides which, some revelations simply never could be made without DNA testing. Like Alice's.

* * *

For more than a week, Alice stared at her 23andMe results. She clicked through the features on the website, scrolling through her genetic cousins, and staring at the little flags in her chromosomes meant to tell her where she hailed from. Gerry's results hadn't come back yet, though they shortly would. Eventually, all the siblings would test, and they'd all turn out to be full siblings, with the same ancestral mix as Alice. But for now, Alice had only her brother Bill's results and her own, and by poking around, she found a feature that purported to show segments along their twenty-three pairs of chromosomes associated with people of Ashkenazi Jewish ancestry. Those portions were blue, and the rest of the chromosomes, represented as bars, were gray. She stared and stared, flipping back and forth between the image of her brother's chromosomes and her own.

Alice did not know what she was looking for, she'd tell me later, but she was stuck on these images. She was doing something she calls "thinking sideways," and over the course of her genetic genealogy research, she'd come to use this technique often. Thinking sideways,

Alice told me, means using intuition to make certain educated leaps, and it's also the trick to solving the crossword puzzles she does every morning. You can't be too literal with crosswords—usually there's a play on words going on, and you have to let your creative mind do some of the work.

So, Alice sat at her computer and clicked and stared, and after a couple of days, all of a sudden, she saw it. The truth was right there, plain as day, in the X chromosomes.

Alice's X chromosomes were represented by a single bar that was part blue and part gray, which meant part ancestry associated with Ashkenazi Jews and part ancestry not associated with Ashkenazi Jews. She knew she had two X's, one from each parent, so seeing all that blue told her only that she'd inherited some genetic material associated with Ashkenazi Jews; it did not tell her which parent was contributing this. But Bill had only one X, given to him by their mother, Alice the elder. He'd inherited his other sex chromosome, his Y, from their dad. And when Alice clicked over to Bill's results, she saw that the bar representing his X was all gray. This meant Alice the elder had contributed no Jewish genetic heritage to her son. It was through their father, Jim, that the mystery lay.

Alice dashed off an email to several siblings, including screengrabs of the images, and walked them through her reasoning. "It is dad's side of the family and I can prove it!" she wrote. "We'll probably never know if it's from our grandmother or grandfather," she added, "but at least the mystery is not quite as big."

But even then, she still didn't know exactly what she was looking at or how big the mystery would become.

* * *

In mid-September 2012, the professor in Illinois known for his skill parsing people's ancestry compositions got back to Alice. His email was filled with numbers—percentages and approximations, and references to "population sets and their fractions," and plus or minus this and that. He attached several images, including a scatterplot, or mathematical diagram, with tiny colored dots meant to indicate different ethnicities.

But the bottom line was this, he wrote: "What you are is 50 percent Jewish and the rest English/Irish/Scots or at least Western European. This is in fact as solid as DNA gets, which is in this case very solid indeed."

This was the moment when Alice knew for sure that the thing she'd discovered by sending in her saliva was way bigger than she'd imagined. The professor's email, combined with Alice's eureka moment in Bill's chromosomes, was clear to her now: It meant that her dad, Jim Collins, was not of Irish descent at all. He was not who he thought he was. The context for his life, and for Alice's, had been altered in a fundamental way.

Alice mulled over and over who her father was—who he'd been and who he'd become. He'd been dead for thirteen years, yet now his life was deconstructed, as if some hand had picked the stitches in one of Alice's beautiful patchwork quilts. He was Irish, *so Irish*, Alice had always thought. "He was raised in an orphanage—he didn't have anything else," she said. "He had his Irish identity." When Jim died, Alice told me, "he wanted an Irish wake. He didn't want people to be sad when he died, and so there were a lot of drunken grandkids." And Gerry, then a cantor at her Catholic church, had sung their father's favorite song, "Danny Boy."

The kids thought their dad looked Irish because he was Irish, and therefore this was what an Irish person looked like. "I grew up thinking Irish people were very little," Gerry told me. Their mom, who was half-Irish, was also short, after all. But Gerry remembered traveling to Ireland in 1990, with her husband and boys, and searching all over the country for her father's face in the crowds. It wasn't there.

"There was nobody that looked like my dad," Gerry said.

As a kid, Alice and Gerry had always thought their dad reminded them of the man Alice called "that famous Irish actor, Peter Falk," the guy who played the low-key, brilliant New York detective Columbo on the TV show. It was lots of things about Falk—the thick New York accent, his self-effacing nature, the social awkwardness coupled with that deeply analytical mind. Plus, Alice thought her dad *looked* like Columbo: the face and the build. It was only in adulthood, when Alice got her unexpected DNA results, that she looked up Peter Falk and learned he was Jewish, born to two people of Eastern European ancestry.

Alice sent the professor's conclusions to some of her siblings. Now she needed the genetic data from their two first cousins more than ever, and the wait was killing her. The kit for her maternal first cousin in Long Island had been sent to the wrong house and went missing for weeks before it was retrieved, while the saliva from her paternal cousin Pete's kit had to be processed twice by the 23andMe lab. Alice and Gerry no longer needed the cousins' results to tell them which side of the family had passed on their link to the Ashkenazi Jewish people—they knew now it was their father. But cousin Pete's DNA could help them understand why their father had never known about this heritage. Could their paternal grandparents, John Josef and Katie, have been Jews who assumed new identities when they came to the United States? Or perhaps the mystery was bigger than John Josef and Katie—perhaps their own ethnic backgrounds had been hidden even from them?

8

SEARCH ANGELS

When CeCe Moore started doing the work she'd eventually become known for, there weren't even words for the things she was doing. She and other citizen-scientists helped popularize the term "genetic genealogy" to describe their work, and Moore started using "unknown parentage" to describe situations in which a person doesn't know the identity of one or both biological parents.

The major genetic testing companies have made all sorts of innovations over the years to make it easier for people to find genetic relatives and to understand where their ancestors may have come from. But many of the techniques for deciphering family connections have come from citizen-scientists, a good number of whom volunteer their time to solve other people's genetic puzzles. Michael Hammer, the University of Arizona geneticist and FamilyTreeDNA scientific advisor, watched this unfold. "In the early days we'd go to these conferences and have workshops and train people on basic genetics," Hammer says. "Over time, because there were folks who were actually getting the data and working with the data, they started asking questions I couldn't answer." He remembers seeing retired physicists and engineers putting their genetic information on spreadsheets and writing their own code to analyze it. Many of these citizen-scientists told me of staying up late to try out new online tools, often getting lost for hours or even days in their research. Roberta Estes talks of these elaborate distractions as "chasing squirrels."

"I have a couple of collaborators I work with on genetic stuff," Estes says. "We'll be working on something and one of them will say, 'Hey, did you see this, did you try this?' and hours later it's three in the morning, we're all exhausted, and someone will say, 'Hey, whatever happened to the thing we were working on?' And someone will say, 'Squirrel. We saw a squirrel.'"

Estes told me that she spent so much time on surname projects traced through Y-DNA data at FamilyTreeDNA that years ago she was able to make a presentation on mutation rates on chromosome alleles among many generations of men born to a common ancestor born in 1495. The rates "differed dramatically from ones that had been published" to that point, she told me, and afterward, scientists attending the conference "came running down the aisles" to her. "This is so exciting!" one told her. "Can we have your data?"

Some genetic genealogists publish peer-reviewed papers on their findings alongside academics, lecture and teach workshops, offer their services to private clients, and act as consultants or unpaid, informal advisors to the DNA testing companies. They publish research in *American Ancestors* magazine explaining how they traced a three-hundred-year-old NPE using thirty-seven-marker Y-DNA tests to distinguish a man from haplogroup subclade E1b1b1 from a family of men with haplogroup subclade R1b1a2. Debbie Kennett publishes scholarly papers on the history of the ancestry testing industry and its social and privacy implications, and gives talks with names like "The Joy of Surnames" and "Mysteries of the *Titanic* Solved by DNA." Blaine Bettinger, an intellectual property attorney with a PhD in biochemistry and molecular biology, crowdsourced and compiled the shared centimorgan data for more than twenty-five thousand known genetic relationships, putting the ranges into a chart to help people decipher how they may be related to each other. He publishes books, runs a subscription help site called DNA-Central with courses, webinars, and a newsletter, and has a blog with posts like "Heteroplasmies and Poly-Cytosine Stretches—An mtDNA Case Study." The International Society of Genetic Genealogy, an influential volunteer-run organization founded in 2005, promotes genetic genealogy, tracks the industry, educates consumers on the science, and maintains an exhaustive wiki. There is no certification for becoming a genetic genealogist, though leading practitioners have collaborated on a set of standards intended to deal with ethical challenges and best practices.

If the genetic detectives can be obsessive, CeCe Moore leads the way, pulling all-nighters on behalf of other people's mysteries, pulling at threads until decades-old secrets unravel. In her work as a "search angel" for those trying to solve genetic mysteries, she's been known to work sixteen-hour days on her beige couch in the Orange County, California, home she shares with her partner and teenage son, staring into her MacBook. In the late 1990s, Moore was a struggling actress, artist, model, and singer, constantly hustling to make a living, when she decided to put a family tree together for her niece as a wedding gift. She considers this the moment when she was bitten by the genealogy bug. But before that, she'd long had an interest in the sciences and, as a young adult, had briefly thought about becoming a population geneticist before settling on music and the theater at the University of Southern California.

Moore started testing herself and family members, eventually getting forty relatives to test. She began answering questions at the International Society of Genetic Genealogy's Yahoo group DNA-NEWBIE, learning as she went along. In 2010, when far fewer than a million Americans had undergone commercial genetic testing, she started a blog series called "Known Relative Studies," to see how patterns of genetic inheritance played out in a real family—her own. "At that time, all we had was computer-simulated data," she says. Moore posted screenshots of various family members' shared genetic material, comparing three generations of her immediate family against a second cousin. It was a new way of approaching genealogy. Rather than looking solely at a paper trail, she could watch the biological trail of her great-grandparents—George, an immigrant from Australia, and Fredrikka, an immigrant from Norway—play out in their descendants, the shared segments of DNA getting chopped up and disappearing over the generations.

Moore told me that when she examined the Google analytics for her blog, *Your Genetic Genealogist*, she was struck by the fact that its readers had IP addresses from places like Harvard and Stanford. After spending so many years being ground down by a field focused on image, she found it gratifying to be recognized for her brain again. It drove home to her how wide-open this new field was—if academics were looking to a citizen-scientist like her for wisdom, she thought, there might truly be a place for her in it. She was in her forties by the time she found her calling, "the thing that I always should have been doing, but it didn't exist."

Eventually, all those years of struggling in the entertainment industry became precisely the background she needed to wind up in front of the cameras again, this time on behalf of science. Moore's talents lie not only in her ability to keep track of details, to multi-task, to follow her squirrelly obsessions up trees, but also in her ability to see patterns in what the DNA is telling her and to translate those patterns, and the science behind them, into explanations that others can understand. Moore has since become a regular fixture on television, the most recognizable face of genetic genealogy. She tells of her exploits solving unusual cases on *20/20*, *Good Morning America*, and *The Dr. Oz Show*. There exists endless B-roll of her typing into her laptop, or scrawling family trees on whiteboards, or strolling with a thoughtful look on her face. Her paid media work has made possible the work she did for years for ordinary people, for which she didn't charge. She and her volunteer administrators, many of whom she helped train, run a Facebook group called DNA Detectives, which has well over 110,000 members, as well as several subsidiary DNA search groups, including one for foundlings, one for the donor-conceived, and one for discussing the emotional ramifications of genetic genealogy. In the past, to find relatives in databases, Moore often advised people to "fish in all four ponds," paying for testing at 23andMe and AncestryDNA, and then uploading their data at FamilyTreeDNA and MyHeritage, which accept transfers. She'd say that so long as she could find a predicted second cousin, she could usually solve a case, and she never knew which company, if any, might have that second cousin. These days, though, with the size of AncestryDNA's database, she simply tells people to upload there; combing all four sites is often not necessary. As recently as 2012, when Alice started her search, "getting a second cousin was winning the DNA lottery," Gaye Sherman Tannenbaum, a retired forensic accountant who's now a DNA search expert, told me. "It's more rare now *not* to get a second cousin match."

When possible, Moore and her fellow angels teach the seekers how to solve cases themselves—"teach a man to fish," the saying goes—but for years Moore pitched in on the most confounding cases herself, giving the seekers her long nights. She helped solve the case of Paul Fronczak, a complicated, years-long mystery involving a man who was abandoned as a baby, and then brought up by the wrong family after being mistaken for *their* baby, who had been kidnapped after birth. In 2015, she and her team cracked the true identity of an amnesiac known as Benjamin Kyle, who'd been discovered unconscious, naked, and injured behind a dumpster over a decade before. And, when DNA testing by Moore's own brother-in-law revealed previously unknown African ancestry, Moore began to dig and eventually traced his direct ancestry back to Madison Hemings, who most historians believe was the child of Thomas Jefferson and his slave, Sally Hemings.

Moore also helped a family uncover why their young adult daughter's DNA test revealed her not to be related to her father—it turned out she was the biological child of a now-dead convicted felon named Thomas Ray Lippert, who in the 1980s and '90s worked at a fertility clinic where they'd sought artificial insemination. Moore believes Lippert, who she says was a prolific sperm donor in the Salt Lake City area, intentionally swapped in his own sperm for that of the father in this and several other cases. The work she does, and what it reveals about human nature, can be disturbing. There is incest more often "than I think anybody could have imagined," she says. She offers resources to the adult children of incest and started a database to study the laws of inheritance in such cases, in order to make better predictions about how seekers' parents may have been related. She hears many stories about the mistreatment of unmarried pregnant women in past decades. "People thought they could get away with everything," she says—and indeed, they did, until genetic genealogy came along. In Moore's view, DNA is an equalizer, a revelatory force with the power to right past wrongs.

I got to know Moore a little through phone conversations over several years, though I met her only once. We were both speaking at a genealogical conference in Waltham, Massachusetts. She was there to give the keynote, as well as a basic primer on using DNA in genealogical research. She relayed so much information that by the end of the first hour, I had filled eleven pages of 8½-by-11-inch paper with cramped handwritten notes. She arrived just before her first talk and left right after her second, and in between, a stream of people waited to introduce themselves to her.

When we spoke—often when Moore was in the car traveling back and forth from various speaking engagements—she always sounded frenetic, her work life and her personal life bleeding together to give the impression that she was never really off. She had a particular kind of drive: Despite her accomplishments, she seemed perpetually dissatisfied. She worked seven days a week, eighty-hour weeks at minimum, for years splitting her time between working on *Finding Your Roots* and helping seekers pro bono. The only time I heard her talk about having free time was after she broke her shoulder on a trip abroad; then we talked for hours because she was having trouble using her computer.

* * *

One of the major forces of genetic genealogy, and a big chunk of the population that search angels like CeCe Moore have helped, is the adoptee community. Unlike Linda Minten, most set out with the knowledge that they're adopted, and use DNA to make an end run around state legislation that in many cases still limits adoptees' access to their original birth certificates. Many of the techniques pioneered to help adoptees find their birth families were created on behalf of, and by, adoptees. In their search for origins are lessons for many other seekers who never expected to find themselves on such quests.

Before the 1990s, most American adoptions were closed, says David Brodzinsky, a professor emeritus of clinical psychology at Rutgers University who spent his career researching psychological issues within the adoption and foster care systems. Now, the majority are open, which means many young adult adoptees grow up having some relationship with their birth parents, or at least knowing something about them. But among those born many decades ago, and among international adoptees to the United States, where openness is less common, gaining access to basic facts about themselves—their own birth names, ancestral backgrounds, the names of the women who bore them—can be incredibly challenging.

"The way we handle adoption in this country still is archaic," says Amy Winn, the president of the American Adoption Congress, which advocates for unrestricted access to original birth certificates, framing it as a basic civil right for all adopted people. Yet with DNA testing, Winn says, "we can go out and find ourselves. It's possible within ten years none of these laws will be barriers to adoptees in search of their origins."

Winn, a Sante Fe, New Mexico, therapist who works on adoption trauma issues, is in her early seventies, and she told me she began searching for her birth parents as a young woman. Like many others, she eventually resorted to hiring a search agency to help her. She was forty-five by the time she gained access to a short letter written by her adoption agency with basic facts about herself. Those three paragraphs with the most meager of biographical details—the fact that her mother was of English and French Canadian descent, and that her maternal grandfather suffered from bad eyesight—meant the world to her. When she spoke to her birth mother for the first time, eighteen months before the woman died, "I heard her voice and I literally felt myself go upright—I had roots to the earth for the first time," Winn said. Friends told her even her voice sounded different after this.

Winn told me the first time she met her half-siblings she felt more comfortable in their presence than she ever had in the home where she was raised. "We are connected to other human beings on this earth—genetically, historically," Winn says. "It's like magnets. I may not know where the other end of the magnet is, but I'm being pulled to it. How can we answer anything about ourselves if we don't know what our roots are, if we don't know who our people are?"

When I spoke with Winn, I was struck by how similar her words were to those of other adoptees. Genetic genealogist and adoptee Gaye

Sherman Tannenbaum told me her experience of ancestry testing was "grounding, it's validating. You've put your stake in the ground—and it's in Belarus or it's in French Canada or it's in New Orleans. 'Wow, my people came from here.'" In the middle of the twentieth century, psychologists coined a term for the sense of disorientation and disconnection that fuels adoptee searches. They called it "genealogical bewilderment."

The desire to know more about one's birth family is normal, says Brodzinsky, and not indicative of a bad relationship between a child and her adoptive parents. "It's part of the universal search for self. It's finding out who you are. It's trying to understand characteristics. We look at our parents and older relatives and we can project ourselves into the future," he says. Searching for family doesn't always mean that an adoptee wants to meet his or her biological parents. But it can mean an adoptee asking questions about where in Russia she comes from, or what her biological mother might have looked like, or why that biological mother couldn't keep her baby. Brodzinsky told me that when he speaks with adoptive parents, they often ask what the chances are that someday their children will search for more information about their biological family. He tells them, "One hundred percent."

In explaining why they search, many adoptees also cite their desire to find out more about their genetic family health histories. They speak of a feeling of frustration and alienation at not being able to answer the most basic questions when they are at the doctor's office. "I just found out heart disease runs in that family," an adoptee named Krista Driver told me months after discovering the man she'd thought of as her genetic father was not. Now, she had learned that four people in her genetic father's family had died of heart disease. "When I go to the doctor, I'm going to ask them to do a full workup," Driver said. "I would never have had that information."

But the interest in medical issues is broader than managing the risk of a discrete disease. Researchers Thomas May and Harold Grotevant have found that adoptees want their genetic family health history because that information informs how they imagine and plan out their lives. The process of seeking out genetic information may serve another purpose as well, May and Grotevant suggest—it may allow adult adoptees, who had no choice in whether to be adopted, or by whom, to exercise their autonomy in making meaning out of it.

* * *

But what about the adoptee's biological parents? How do they feel about being found? Brodzinsky told me social science research suggests adoptive parents, adoptees, and biological parents all tend to fare better, psychologically speaking, under open adoptions—and that, historically, research into closed adoptions in which children have managed to track down their biological mothers shows that most of those mothers are ultimately glad to be found. But this is not a guarantee, and the word "ultimately" is key—seekers are on their own timeline, and it may not be the same for those they seek. It is a different thing to find someone than to be the one who is found, and it can take time to come around to it. "We tell people if they get rejected . . . wait," says Karin Corbeil, a co-founder of DNAAdoption.org, a nonprofit that helped pioneer genetic genealogy methods to help adoptees. "Many times, the shock is so overwhelming to people." This can lead to situations in which a seeker's interests can seem to be in conflict with the person they seek at the most delicate time—at the very beginning of a relationship that has the potential to be intense and intimate.

"What we're seeing now is a more direct conflict between the privacy rights of genetic parents who have given their children up for adoption and adoptees," says bioethicist Thomas May, who is himself an adoptive father. Technology is "advancing to the point that it's forcing us to rethink the very manner in which we adjudicate balancing these interests." Should parents have an expectation of privacy if they relinquished their children decades ago when the culture was radically different, when psychologists said the best thing for a child was not to know his or her biological family?

I have heard of cases in which birth mothers did not want to be found because their children were conceived by rape or incest, or where such traumas were the explanation for misattributed paternity. Brianne Kirkpatrick, a genetic counselor who works with those affected by genetic surprises, told me about one seeker she worked with: "The woman said, 'I told my mom that the man she told me was my biological father wasn't my biological father. And she had this look come over her face, and she had to tell me her uncle had raped her.'"

Because of the shared nature of genetic information, it's not always clear whose desires should be honored when it comes to the question of relatedness. In 2017, AncestryDNA made it possible for consumers to opt out of seeing or being seen on relative match lists. It framed the decision as part of its commitment to customers' privacy and control, but the comments section underneath the company's blogged announcement was riddled with dismayed messages, and one man accused the company of "discriminating" against adoptees.

Concerns about the privacy rights of mothers and fathers are why Roberta Estes, who consults to people who want to better understand their genetic results, told me she doesn't take on cases in which people are searching for their genetic parents. Estes doesn't object to people knowing the identities of their genetic kin; indeed, she wanted badly to know her brother, Dave, and she did not see her letter to him as likely posing an unwelcome disruption to his life. But, she said, it was impossible to know whether birth parents would welcome being found by their genetic children—if only because they'd never told anyone about what was seen back then as the sin of having a child outside of marriage. Estes worried that some practices she'd heard about—tracking down birth mothers through social media and sending registered letters to siblings—could be experienced as invasive and meant that "whether the mother chooses to or not, she's outed because of the way the child chooses to find her." The only way to know if a mother welcomed her child's contact was by contacting her, and the act of contacting in and of itself changed the mother's life.

"I'm not saying I'm opposed to it," says Estes, who refers cases such as those to DNAAdoption. "I'm saying I have a concern and I don't do that work."

DNA testing poses questions that people in the adoptee community have thought about for far longer than such testing has been around. Will my birth mother/father/half-sister be happy to know me? And, more broadly, what is that person to me? What do we mean when we speak of "family"? How much does genetics get to tell us about who we are?

Some people I spoke with, who did not know that one or both of their parents were not genetically related to them till they tested, told me that in retrospect, they always sensed it. Dani Shapiro, author of the memoir *Inheritance*, about her experience discovering by DNA testing that she was conceived by donor sperm, has said she thinks she always wondered, obsessed as she was for years with writing about family secrets and the search for self—which is not to say that, in discovering the deeply meaningful truth about her biology, the man who raised her was any less her father. He "loved me into being," she writes.

Stories like that of Shapiro's sixth sense raise the question: Could there be a biological imperative to seeking out one's own genetic family? This is what evolutionary biologist Jerry Coyne, author of *Why Evolution Is True*, suggested to me. During most of human history, "kinship was of real importance to humans because we lived in small groups and we knew each other and it was important to know who our kids were," Coyne says. "During this period of time we developed—not consciously but evolutionarily—a way of sussing out who shares our genes and who doesn't." We give special preference to our own children, as do other species throughout the animal kingdom, he says. Natural selection may foster "mate guarding" and sexual jealousy among males, who want to know that their children are biologically related to them. "It becomes a matter of special interest to humans to know who shares their genes," Coyne says, and it may be that adoptees' urge to seek out kin is based, in part, on this evolutionary urge we all share.

But it can be difficult to tease out biological impulses from cultural influences. Other seekers who discovered biological surprises through testing have said they had no inkling those surprises were coming. And I've also spoken with people who believed—hoped, even—that they were not related to their genetic families, yet testing confirmed

that they were. Ours is a culture in which we make family many different ways—through traditional conception, through adoption, through
donors, through stepparents—and in my interviews, I found people
defined their families with nuance and deliberateness. Kimberly Leighton, a professor of philosophy at Georgetown University and herself
an adoptee, is deeply critical of the term "genealogical bewilderment,"
which she says reflects an old-fashioned notion that heredity is key to
self-understanding, and an assumption that the best kind of family is a
genetic one. The real cause of distress among people who are adopted
or donor-conceived and don't know their genetic families, Leighton
argues, may be a response to a culture that values heredity as the true
measure of family and identity. This doesn't mean that adoptees and
donor-conceived people may not want to know more about their genetic
parents; it simply means that wanting to know may have much to do
"with how we value genetics," she says.

"I really think it's the quality of the relationship that matters, and
genetics plays a very small part in that," says Thomas H. Murray, who
has written extensively on adoption and genetic bonds. Parenthood is
"propinquity. It's being with someone hour after hour, day after day,
when they're crying as an infant, when they're sick and need your care."
But if genetics pales in comparison to a lifetime of interactions, Murray
says, it's also true that there's a long history of belief that inheritance
matters, "partly because cultures cluster along with genetics, and culture matters—cultures are often what define alliances and enmities."
The questions of what makes a family are as old as families themselves;
only now, in this new era, many more families than before are grappling with them. In one of his books, Murray points to a heartbreaking
declamation from a man in ancient Rome whose foster son had been
reclaimed by his biological parents. When that child was away at war,
the man asked the court, "Whose was he? Was it not I who hung from
the city walls in suspense? . . . Who bandaged his wounds when he
returned? Who washed away the blood?"

I asked Harvard Law School professor I. Glenn Cohen how US law
defines family and whether that has been changing in recent decades.

He told me that, starting in the 1990s, courts have dealt with new kinds of families—many of them with children conceived by assisted reproductive technologies like donor insemination, egg donation, and surrogacy—by placing them into the context of an old-fashioned nuclear family unit. This has at once reinforced the traditional idea that the nuclear unit is the model for the American family, and allowed for a reinterpretation of the family to accommodate gay couples and single parents. In doing so, Cohen said, courts were also reinforcing a social definition of family over a genetic one. They did not want to encourage "strangers" stepping into the nuclear family—even if those strangers were genetic relations.

It remains to be seen, Cohen said, whether consumer genetics will push the balance in the other direction, toward a view that reinforces genetic essentialism; he does not think these tests will. And he added that discoveries of paternity as a result of genetic testing are unlikely to lead to major legal issues around inheritance or child support. Parents are under no obligation to provide for their adult children in their wills, so the adult child of an NPE situation would have a hard time making a claim on his or her genetic parent's estate. As for minors discovering their genetic parents via testing, "most states have a rule that paternity essentially becomes final after a period of time unless challenged," Cohen said, and besides which, sperm donors are generally considered not to have paternal obligations when they donate anonymously through sperm banks.

In recounting the story of finding his birth mother, adoptee Tim Monti-Wohlpart, the national legislative chair for the American Adoption Congress, told me that attempting to answer the question of nature versus nurture in definitions of family and how it shapes us is a fool's errand. So are assumptions that an adoptee could have room in his heart for only one mom. That represents a failure of the words we have for such relationships, not a failure of the heart, because both things matter—for adoptees and for parents. "People do a lot to have children that are genetically related to them," says Ellen Wright Clayton, who teaches law, pediatrics, and health policy at Vanderbilt University.

"But I believe down to the tips of my toes that family is about relationships, about the social part, however it's formed. And I think people get to choose."

* * *

There is something uniquely American about the growing mass of seekers that has arisen over the last decade. Genetic genealogist Debbie Kennett, who tracks the industry, estimates about 80 percent of the market for DNA testing is in the United States. The next biggest markets are in the United Kingdom, Ireland, Australia, New Zealand, Canada, Finland, and Scandinavia.

Europe's market is seen as several years behind the US market because of a complex tapestry of policy, pragmatism, and culture. In general, says David Nicholson of UK-based Living DNA, Europeans are more concerned than Americans with matters of privacy and security. The regulatory landscape for direct-to-consumer DNA testing abroad is fragmented, with some countries, including France and Germany, imposing many more restrictions on how tests are sold. Andelka Phillips, a legal academic, says the European Union's General Data Protection Regulation, which went into effect in 2018, as well as additional EU regulations, may force more transparency onto genetic testing companies and may even restrict how some American companies operate in Europe.

There are also cultural reasons for this reluctance. In some parts of Europe, there are concerns that using genetic data to search for ethnic differences is reminiscent of the rise of eugenics in Nazi Germany. "I have some German friends who think that the idea of testing to see what percentage of ancestral origins you are is the *opposite* of what you should be doing," CeCe Moore told me. Besides which, says Nicholson, "people in genealogy across Europe generally have their records going back thousands of years anyway, so how's DNA going to help them?"

There's long been an idea that the New World is where immigrants remake themselves free of the past, yet it's precisely because Americans are so divorced from our histories—because we lack the continuity and the ancestral knowledge that comes from generations living in the same place, in a largely homogeneous nation—that many of us wonder where we came from. We are intrigued by our roots, curious about all the things that took place before our ancestors traveled across the ocean, or across the American continent, and settled here. We want to know the specific bits and pieces that make us *us*.

A media industry has arisen around this cultural curiosity. The Harvard scholar Henry Louis Gates Jr. has been at the forefront of this, hosting *Finding Your Roots* as well as other PBS specials combining genetics and genealogy. Ancestry sponsors *Who Do You Think You Are?*, which has run for many seasons on NBC and TLC and also traces celebrity genealogies. A growing number of people are taking heritage trips to the countries they discover they hail from when they take a genetic test, and the DNA companies are getting in on the action. AncestryDNA teamed up with a travel outfit to offer guided ancestry tours to countries including Italy, Ireland, and Germany; and 23andMe got together with Airbnb to showcase "homes and heritage travel experiences based on a customer's DNA."

And on YouTube, the "ethnicity reveal" has become a thing. In thumbnails, the people who post these invariably have shocked looks on their faces and tend to employ a lot of exclamation marks. One woman—who titled her video "Reading My Ancestry DNA Results! I Been Lied to My Entire Life!"—filmed herself discovering that "my dad isn't really my dad." She pauses the video to call her dad, then gets back in front of the camera. "He told me that he always had an inkling," she says, exhaling slowly and trying not to cry. So far, it's garnered more than 1.5 million views.

Over the last few years, I've spoken or emailed with more than four hundred people about their DNA testing experiences. They told me about how they were converted into seekers, about the wormhole that is the search for the past and for the self through genetic genealogy. They told me about the many tools they used, about testing in multiple places, and about using all sorts of third-party inventions, including browser extensions, software, and tools at sites like GEDmatch and DNA Painter to sort, compare, and predict relationships, interpret DNA segment data, keep track of notes, and filter matches.

Some seekers subscribe to newsletters like biologist and genetic genealogist Leah Larkin's *The DNA Geek*, which chronicles techniques like the McGuire Method and the Leeds Method (to "help you depict and organize your DNA matches, respectively"). Online, seekers commiserate over their brick walls, and discuss their maternal and paternal haplogroups. Some talk about pulling all-nighters, about waking in the middle of the night to use the Ancestry app on their phones, and working it until their phones die. Their spouses call themselves genealogy widows and widowers. There's something terribly addictive about this hobby and the questions that it poses: How much do you know about yourself and your family? How much *can* you know? ("I have been called a crack dealer," FamilyTreeDNA's Janine Cloud told an audience at a genealogy conference, and everyone got it.) The seekers talk about checking for relative matches several times a day and about the kismet of noticing family surnames on street signs. One told me of a brick wall so befuddling and engrossing that at one point an entire wall down a hallway of her home was covered in the names of ancestors and relatives on sticky notes, which she spent her evenings and weekends researching and rearranging. Seekers are stalkers of the living and the dead.

In my experience, both seekers and search angels are more often female. This makes sense: Brodzinsky told me that among adoptees, it's the women who are more often seeking to connect with their biological families. Perhaps it's in part because of the way girls are socialized to express their emotions and "to actualize a search based upon those feelings," he says, and perhaps pregnancy, childbirth, and motherhood form a strong impetus for women to connect with their own biological mothers.

Laurie Pratt, who became a search angel after discovering she was the product of an NPE, told me that after working in genetic genealogy awhile, she started to develop an instinct for it. Piecing together people's family trees draws on all sorts of knowledge—knowledge of history, of psychology, of geography. For example, if a person wants to figure out the identity of a genetic relative who goes only by a nickname on Ancestry, and offers no family tree, she recommends searching that nickname with a variety of email address suffixes, like @gmail, @hotmail, @yahoo. "People are creatures of habit," she says, so if you're calling yourself, say, ShyBunnies123 at Ancestry, chances are you're using ShyBunnies123 someplace else. Pratt says reading the obituary of her genetic father's brother helped her get a feel for the family he came from, the way it saw itself and the things it valued. As a search angel, knowing the landscape of Orange County, California, helped her trace the identity of a baby abandoned in a supermarket, because she knew, based on where certain people lived, how easily they could have reached that supermarket by way of a certain freeway. And she could see, based on clues she was finding online, a pattern of troubles within a family that suggested certain people might be desperate enough to leave a child.

"As people, we're the sum of other people's choices: the kinds of jobs they take, where they choose to live, the stories they have," Laurie says. Whether through social media or hundred-year-old records, a genealogist can start to see patterns: Is a family upwardly or downwardly mobile? Have they lost a child? How does dysfunction beget dysfunction? These are just clues, not certainties, but they can help make the work of a laptop detective more efficient, telling her where to invest her efforts. Laurie told me her fastest solve involved a woman who didn't know the identity of her father, but had a big clue of a match: a close DNA relative, though exactly how close it was hard to tell. Laurie pieced together the DNA relative's tree using records online, figured out who his grandmother must be, deduced that the relative was actually a half-brother to her seeker, and from that discovered the identity of the seeker's father. It took about twenty-five minutes.

Pratt can use relative matching in her genetic detective work because it works so well. While the science of ethnicity predictions as carried out by DNA testing companies has been heavily criticized as not as precise or accurate as is implied by those pie charts of 3.1% *this* and 26.2% *that*, relative matching relies on a different approach.

"Matches are incredibly reliable," says geneticist Barry Starr, Ancestry's director of scientific communications. "What you're doing with relatives is you are looking for large chunks of DNA that are exactly the same between two people. Unless you're a close relative, you're not going to show identical DNA."

There are cases where relative matching isn't perfect. Predicting just *how* two people are related is a challenge; a company can tell you the amount of genetic material shared but, in many cases, can't tell you precisely what the relationship is. A few years back, MyHeritage DNA updated customers on improvements meant to address some cases of "false positives," in which they were overestimating the shared segments of very distant relatives, making them look closer than they were. And in one unusual case, genetic counselor Brianne Kirkpatrick told me a woman reached out to her after 23andMe predicted she had a "half-sister." In fact, the mystery "half-sister" turned out to be a first cousin who shared an exceptionally large amount of genetic material with Kirkpatrick's client, a fact she and the client were able to verify through additional testing of another family member. "She's still trying to recover her thinking around her father because for a month, she thought of him as a man who cheated on his wife," Kirkpatrick told me shortly afterward. "She said she's having a hard time shaking it."

Because of the way genetic material is shuffled and passed down randomly, with chunks reconfigured, trimmed, and lost in each generation, you may not share detectable DNA with, say, a fourth or fifth cousin or beyond, which means they won't show up for you on your list of matches. In some cases, this is even true for third cousins. As computational biologist Yaniv Erlich has pointed out, "Even if you are connected to a person who came over on the *Mayflower* in 1620 (about thirteen generations ago), chances are that you did not inherit any of this person's DNA. If we could do a genetic test, you and your *Mayflower* relative would look like two unrelated people." In other words, even though someone shows up on your genealogical tree, that doesn't necessarily mean they'll be in your genetic tree. (Science journalist Carl Zimmer points out that this throws cold water on our genetic essentialism, shattering whatever illusions we might have about inheriting some important ancestor's royal blood, or another ancestor's ignominy, via their DNA.) But the system and science of relative matching through autosomal DNA almost always works quite well for the research Laurie Pratt and other search angels do, relying on multiple cousins, ideally as close as possible, coupled with traditional genealogy techniques, to understand precisely how people are related.

Though, in truth, Laurie told me, by 2019 she was rarely doing search angel work anymore, because seekers were finding so many close relative matches that they could easily unravel their own cases in a day with just a bit of guidance. It turns out, seeking is increasingly the easy part. It's the finding that can be complicated.

* * *

What happens when the seeker finds? When an adoptee or another seeker has identified her long-lost kin, what does she do with that information? No chart or browser extension can answer this question, nor can algorithms predict how the people she's identified will react to being found. This is when some seekers prowl the Internet, looking for clues into whether they want to know these strangers in all but blood. It is, of course, breathtakingly easy to figure out your father's career on LinkedIn, his political leanings on Facebook. Social media makes it possible—once you've discovered *who* your father is—to learn *what* he is.

The question of what to do with genetic information once you have it has inspired its own class of advice. Not every seeker reaches out. "Who am I to upend your whole life with this knowledge?" one woman said about her decision not to tell her biological father of her existence. But many have a desire for contact. And then: How should a seeker contact a genetic relative, if she decides to, and in what fashion, and what should she do if she meets with silence? With outright rejection? Some people reach out through a DNA testing company's messaging system or by email, whereas others turn to old-fashioned snail-mail letters even as the rest of the world is moving away from them. (No matter how cutting-edge the technology for identifying a relative, there's nothing like the gravitas of a physical piece of paper for contacting him.) Some genetic genealogists advise first taking screenshots of family trees and other materials proving their match, in case a relative decides to delete his or her account after being contacted.

There are "two main schools of thought," says Stephanie Talley, who found her sperm donor father through genetic testing and, as an administrator for a Facebook group, sometimes counsels other donor-conceived people about how to write that first letter to a donor. The approach she suggests is "initially being very light and friendly and maybe not giving them everything right up front. Just saying, 'I've found this connection. Are you interested in getting in touch? This is who I am.'" The key is "making yourself seem like somebody who is friendly, not terribly dramatic . . . because it's a lot. It's heavy anyway."

The other school of thought, Talley says, is for the seeker to offer lots of information about herself because "what if this is the only chance I have, and they tell me 'Don't contact me again,' and I never got to say what I wanted to say?" Talley says she understands this instinct, but putting everything into a letter can make a seeker seem "a little stalkery" and produce "this really high-pressure letter for the recipient." Besides, holding back allows for more conversation fodder later. "Why give up everything that might be bait for them to meet you?"

Those just finding their parents in adulthood are often in a defensive crouch, scared of rejection, and this perspective can seep into their letters, Talley says. She understands it; she tells fellow seekers that if they don't throw up from the stress and anxiety of these early interactions with biological family, they're doing better than she did. "It feels like the most important job interview ever."

And if the seeker's parent does reject her, and she's laid her heart out on the screen, how will she feel then? Because, in the end, these may be genetic parents that seekers are reaching out to, but they are also strangers—strangers who can choose, in ways most parents can't or don't, whether to have relationships with their children. Talley does not want fellow seekers to feel like they made themselves vulnerable to a stranger who then breaks their heart. She does not want them to regret saying something—or, for that matter, *not* saying something. So, the guideline she gives seekers is this: Imagine the worst: the rejection, the silence, whatever it is. Then write a letter you can live with if you get that response.

Sometimes, in the search for family, a whole lot of clues can add up to the wrong answer. One woman, who discovered she was donor-conceived via 23andMe, told the magazine of the American Psychological Association that early on in her search she mistook the wrong man for her biological father, "resulting in some uncomfortable interactions." And sometimes, getting it right feels like a mixed blessing. An adoptee I'll call Lisa, who tracked down her mother as an adult, told me their relationship could be strange and awkward, akin to the early stage of dating, with the initial nervousness, the testing of the waters, the struggle to set expectations. When they first talked, her mother asked, "Do you hate me?" Lisa did not hate her mother for relinquishing custody; indeed, she was grateful her mom had done it. But now, "If I don't call . . . soon enough, often enough, she gets mad," Lisa wrote in one email. "I just can't give her as much as she wants."

Laurie Pratt never did hear back from her biological father. "I want to see him in person," she told me one summer day. By then, it had been about a year and a half since she'd first contacted him and he'd deleted his test. She respected his boundaries and did not contact him after that single follow-up letter. She told me she could understand how her existence complicated his life and why he might not want to know her. But she continued to get to know cousins from that side of the family; they were her cousins, too. And, after hearing from another seeker who did something similar, she was mulling the idea that she might one day travel to her father's area and look for an opportunity to discreetly view him. After all, he lived in the world. He went to public events. "I just think it's weird to say I've never seen my father in real life," she told me.

In 2018, a group of Canadian third cousins on Laurie's father's side contacted Laurie after reading her blog, to ask if she wanted to come to a family reunion. So she and her husband flew up to their rural farming community, and a cousin showed her the house where Laurie's great-great-grandmother had once lived. She came back feeling empowered by the gift they'd given her—the gift of knowing her roots. And she realized something: For years, she had walked around genealogical conferences with a name tag identifying herself by her maiden name, Laurie Pratt, which was how she was known in genealogical circles. She'd fielded questions about her last name—"Are you related to Enoch Pratt?" "What about the Pratts of Springfield?"—and felt like an imposter, because she wasn't actually a Pratt by biology. Now, she decided to legally change her name, hyphenating her husband's last name with her biological father's last name, in order to claim her identity as someone descended from his genetic line.

"I'm not out to anger him, and I certainly don't think he'd be like, 'Hooray!' but this is reality," Laurie Pratt McBriarty-Sisk told me. "I just want to deal with reality."

9

27 PERCENT ASIA CENTRAL

Among the theories that Alice and Gerry entertained in the early months of their investigations were several that were historically intriguing. Alice wondered whether her grandparents John Josef and Katie might have been Jews who somehow claimed Irish identities when they emigrated to America. Or perhaps the explanation was further back—maybe her grandparents were Irish Jews who'd hidden their religion (or had it hidden *from* them) all their lives.

These ideas seemed far-fetched, but they were ways of explaining something that seemed unexplainable. Besides, history is far-fetched. There are groups of people who've survived by passing. These include the converso Jews in Spain and Portugal who were forced to convert to Catholicism, some of whom continued to practice their faith in secret; and light-skinned African Americans who passed as white in America, escaping rampant discrimination at the steep cost of leaving their families and communities. These secrets were in the headlines long before consumer genomics arrived. Madeleine Albright did not know her parents had been Jewish till a *Washington Post* reporter uncovered it in 1997, when the secretary of state was fifty-nine years old. John Kerry did not know his paternal grandparents were born Jewish until adulthood. Genetic testing has revealed many cases like that of CeCe Moore's brother-in-law, who did not know about his African ancestry until he spit into a tube.

Recreational genetic testing can suggest ancestral homelands and migration patterns that point to a family's true history, hidden for decades or centuries, prompting testers to reexamine not only their own families but their beliefs about race and ethnicity, too. Some seekers are caught entirely unaware when they make this discovery through DNA, whereas others seek out testing because they've long had questions. In the face of those findings, they must come up with their own definitions of ethnic identity, informed by what they were told as children as well as by things they were pointedly never told. It forces a reckoning with history, with the things their ancestors did in order to be able to pass on their genes to them.

* * *

Rosario Castronovo grew up without much in the way of cultural identity, but he clung to his mother's story that she had Sicilian heritage. At twenty-one, when he decided to legally change his name to distance himself from a father he describes as abusive, he chose Italian first and last names in a nod to his mother's culture. He'd heard "Castronovo" meant "new castle," and that's what he believed he was building—a new life, a new identity. Nobody in his family stopped him, though he'd later learn that many of them knew this was a fable.

Come to think of it, a lot of the decisions Rosario made as a younger man were nods to a heritage he'd later learn he could not claim. He joined the Catholic church as an adult and was baptized. He studied and began singing opera. He proposed to his Italian American girlfriend after flying her to Italy for the millennium. He was "trying to define who I was." Many of us have family stories, memories, holidays, habits, and language to assist in the constructions of our ethnic identities. Rosario filled in the blank parts with what he thought it meant to be Italian. He was "searching," he'd say later. "What was going to make me *me?*"

It was when he was about to be married, in the early 2000s, that he decided to find out more about the background of his mother, an orphan raised in foster care. Rosario wanted children, and he imagined that one day they'd ask where they came from, just as he'd asked his own parents. He wanted to be able to give them an answer.

In the town hall in the small town in Vermont where his mother had grown up, Rosario found not one but three birth certificates for her, and they were rife with redactions. Strange. One listed the race of his mother's father as "negro." Rosario was mystified. Perhaps his grandfather had been a dark-skinned Sicilian mistaken for a black man. But more likely, he thought, putting the information together with old census reports that sometimes named his mom's paternal side as black and sometimes as white, the birth certificate was right, and there were important, fundamental facts he did not know about himself and his mother. He did not go to his mother yet because he wanted more evidence before he rocked her world. *Maybe she doesn't know*, he thought.

Over the next few years, Rosario used genealogy skills to research his mom's side, but there were details he could not fill in from the paper trail alone. So, he embarked on genetic testing. He would eventually learn that he is about 18 percent sub-Saharan African, as well as smaller amounts Native American and Asian, all through his mother's side. His elderly mother was uneasy when he went to her with the results of an early autosomal DNA test he'd persuaded her to take.

"I said, 'Did you know?' and she said, 'Yes,'" Rosario says. She said, "I didn't tell you because I didn't want you to go through what I went through." She told him about her difficult life in foster homes, about her foster father directing a racial slur at her and her little brother, about her brother being sent away to a boys' home because, she understood, "he had darker skin."

And Rosario began to learn about the history his mother had wanted to protect him from. He researched the small town in Vermont his mom came from, traced how the slate quarries and mills attracted black Americans, traced how the black families in the town peaked toward the turn of the twentieth century and then began to disappear. He learned of the death of his thirty-year-old great-uncle in what a 1930 newspaper article called a "fire of mysterious origin," when many black families had left the town, and wondered about the circumstances of that fire.

And eventually, he came to understand some of the context for why his mother had lived out her childhood as an orphan, even as her parents were both living nearby. He learned that his grandmother was white and married to someone else when she had two children with Rosario's black grandfather. Looking in archives, he discovered his grandparents had both been sentenced for adultery in the 1940s, though only his grandfather served time, going in when the two children they'd had together were small. Rosario did historical research and wondered if his grandfather's race and his relationship with a white woman were the real reasons he sat in jail. Part of the judge's ruling in Rosario's grandmother's case was that she no longer "make her residence" with his grandfather. She eventually divorced her white husband and married someone else, and Rosario's mother had only the haziest memories of seeing her biological parents as a young child. It was as if her family had never existed.

* * *

You can't tackle questions of ethnic identity without first considering what the sciences can teach us about the concept of race. In recent years, as research into human genetic difference has advanced, discussions about how to define race and ethnicity have become a kind of meme of American culture, hashed out in the pages of the *New York Times* and *The Atlantic* and on Twitter before eventually deteriorating into scientific racism in the Internet's darkest places.

Those of us who don't study population genetics or anthropology for a living often come into contact with this debate, implicitly at least, via at-home genetic testing, and those little pie charts that purport to be able to tell consumers whether or not to root for Germany in the World Cup. Around the 2018 Winter Olympics, AncestryDNA aired one ad that implied a creepy kind of genetic essentialism. It showed a figure skater majestically crossing the ice, while a voice-over mused on what made her so good, conflating certain cultural stereotypes with biology. An "ethnicity" pie chart popped up in a corner of the screen, suggesting it was, perhaps, her 48 percent "Scandinavia" that accounted for her "precision," her 27 percent "Asia Central" that accounted for her "grace," and her 21 percent "Great Britain" that explained her "drive."

"One of the biggest criticisms of this whole industry . . . particularly in the social sciences, is that the existence of these tests is going to
make people think that race is buried in your genes, and reinforce this
essentialist view," says sociologist Wendy Roth, who studies race and
genomics. As the early history of genetics demonstrates, this is more
than an academic concern. As Roth put it in a lecture some years back:
"The view that certain groups are essentially or genetically different can
quickly morph into the view that they are essentially inferior, and that
has been used to justify oppression, forced sterilization, eugenics, and
even genocide."

What does the science show? "If you do genetic analysis, you actually see that we are a mosaic of heritage, and we come from different
places in the world, and that racism has no scientific basis," computational biologist and MyHeritage chief science officer Yaniv Erlich told
me. Erlich said he sometimes wonders how history would have played
out if modern genetic tools had existed a hundred years ago. On the
one hand, the power to determine someone's ancestry would have been
incredibly dangerous in the age of eugenics. On the other hand, he says,
perhaps greater scientific understanding would have allowed people to
see that eugenics "is bullshit."

The science shows that for most human traits shaped by genetics, there are more differences within population groups than between
them. And, at the risk of crushing the hopes and dreams of neo-Nazis,
the notion of racial "purity" is bunk, given the genetic overlap between
different populations and the fact that most genetic variation is "dis-
tributed as a gradient," as the American Society of Human Genetics
puts it, so there's no clear cutoff between one population group and
another. Over the course of history, "human populations appear to have
repeatedly split, merged, and interbred," University of California, Berkeley, bioengineering professor Ian Holmes writes. As a 2018 piece in
National Geographic pointed out, there's more genetic diversity among
people in Africa than on all the other continents combined because
Africa is where we all began and where human beings have lived the

longest—underscoring the point one geneticist made that "there is no homogeneous African race."

The concept of race as distilled into popular culture is the product not only of scientific research but of history, culture, and politics as well—the product, in other words, of flawed human beings. Columbia University sociologist Alondra Nelson has described the concept of race as "a way to sort human communities in such a way as to justify social inequality; this sorting is neither natural nor inevitable."

"Deciding which characteristics matter—whether it's skin color or eye color or height or parentage—that is a social process and it's something that different societies do differently," Roth observes, giving as an example a poet who came from the Dominican Republic to America in the 1960s. Back home, the woman had been considered "indio" or "trigueño," "intermediate categories that are part of a wide continuum" of racial terms. In the United States, where culture is influenced by racial purity laws like the one-drop rule that historically defined someone's race on the basis of a single sub-Saharan African ancestor, the woman was "black." The woman's biology was the same, but her racial definition, and all that came with it, shifted as her culture did.

But if race is, as Harvard geneticist David Reich puts it, "fundamentally a social category" rather than a biological one, it is nonetheless quite real as a lived-and-felt experience for Americans. And so is racism. Among social scientists, there's a good deal of concern over how to talk about the research that scientists are doing into genetic differences among populations. They are careful about the language because they know the power of words like "race," and the assumptions embedded in it, and because they know full well how the history of genetics is intertwined with racism and anti-Semitism. As if to underscore this deeply troubling entanglement, James Watson, the co-discoverer of the structure of DNA, has repeatedly asserted that black people are innately less intelligent than white people, and has linked skin color with sex drive. Watson "fell prey to the same temptation to which the old Progressive eugenicists had succumbed: he let genetic determinism amplify his prejudices and biases," writes historian Nathaniel Comfort. "It is, perhaps, an occupational hazard of those who think about genes too much."

So, how should scientists talk about the real differences among populations that they see? Because while the concept of race morphs from society to society, privileging some groups at the expense of others, Reich writes that there are differences, on average, among populations for a number of traits. "We now know that genetic factors help explain why northern Europeans are taller on average than southern Europeans, why multiple sclerosis is more common in European Americans than in African Americans, and why the reverse is true for end-stage kidney disease," Reich wrote in a 2018 *New York Times* op-ed tied to his book, *Who We Are and How We Got Here*, on new findings from the study of ancient DNA, many of which have come from his lab.

And understanding these differences is important not only for our understanding of disease, evolutionary biologist Jerry Coyne told me. "If there were no genetic differences between groups, it would not be possible to retrace this history . . . to know how we spread across the planet and where we came from," Coyne says.

That discussions of population differences have been misconstrued to justify bigotry is deeply problematic, Reich writes. But also problematic, he argues, is the fact that some scientists are suggesting there are no population differences for fear of having their findings misunderstood. "Arguing that no substantial differences among human populations are possible will only invite the racist misuse of genetics that we wish to avoid," he argues, by causing laypeople to feel misled and patronized. Reich believes the gap created by the reticence of the scientific community is being filled by people like Watson and science journalist Nicholas Wade, whose 2014 book *A Troublesome Inheritance* was denounced by more than a hundred population geneticists and evolutionary biologists, including Reich himself, as filled with racist stereotypes. Also filling the gap: "race realists" (modern-day proponents of old-fashioned scientific racism) and other white supremacists.

"While race may be a social construct, differences in genetic ancestry that happen to correlate to many of today's racial constructs are real," Reich wrote in his 2018 op-ed, provoking fierce debate and underscoring the difficulty scholars are having agreeing on the very language for this discussion. In *The Atlantic*, UC Berkeley's Ian Holmes wrote that while Reich had done a good job of describing "race's complex relationship to ancestry," his op-ed fell down when it used the word "races," albeit in quotes, calling it "a loaded term for scientific outreach." In *BuzzFeed*, more than sixty scholars, mainly social scientists, published an open letter denouncing Reich's op-ed, and elsewhere Joan Donovan, a sociologist who has studied white supremacists' use of DNA ancestry testing, described his words as "biological reductionism." Reich stressed to me that this was the very opposite of what he meant.

At a moment when so many discoveries are being made into human genetic diversity, some scholars are deeply concerned about what happens to genetics research when it gets distilled and simplified by the culture at large, not to mention how assumptions shape the science itself. They know that now, as in the eugenics era, human beings looking for patterns will see what they expect to see. They know history, and they fear its repetition.

* * *

The criticisms of ethnicity estimates are manifold. For one, there's the use of the word "ethnicity," which is not quite right, since there's a distinction between one's ethnicity and where one's people came from historically. Ethnicity is in good part about "culture and tradition and self-identity," CeCe Moore told me—things that you cannot measure with DNA. "Someone can have a cultural ethnicity without the corresponding scientific basis for it in their genome."

Anita Foeman, who studies perceptions of race and ethnicity in genetic testing at West Chester University of Pennsylvania, told me about one of her subjects, a young woman of Chinese descent raised by a German American family. When her boyfriend's parents inquired into her ethnic background, she told them she was German. They laughed, thinking she was joking. How strange, the young woman recounted, to have people look at you and tell you you're something different from what you are. "What does a DNA test tell you when it's 100 percent not how you see yourself?" Foeman asks. "What makes you German? And can she own 'Chinese' even though she knows nothing about that culture?"

Yet the largest of the consumer genomics companies, AncestryDNA, has popularized the term "ethnicity," and it has taken off in part perhaps because of what cultural psychologist Steven J. Heine calls our "essentialist thinking," which encourages us to conflate genes with identity. Besides which, other, more precise terms aren't consumer-friendly. "Biogeographical ancestry is such a mouthful," says genetic genealogist Debbie Kennett.

The other major criticism of ethnicity estimates is that, while they are billed as estimates, they are presented as offering a degree of precision that companies cannot actually deliver on, for reasons having to do with their imperfect methods for interpreting the data in our genes, as well as the complications of history. They are "a young and developing science," a MyHeritage DNA speaker told a genealogy conference, reiterating what anyone who's had their results "updated" by a company knows. A number of genetic genealogists have blogged and spoken about the differing ancestry estimates they've gotten from various companies. "I had almost 50 percent Scandinavian when I took Ancestry's first test," Roberta Estes told me. "You know how much Scandinavian I have now, several iterations of the test later? Guess what: I don't have any Scandinavian."

"It's dead easy to tell continents apart," Barry Starr, Ancestry's director of scientific communications, told me. The DNA of someone whose ancestry is from Asia looks quite different from that of someone from Europe or Africa. The ancestral estimates of different companies will generally agree when it comes to such broad geographic areas. But because continent-level predictions are usually unsatisfying to people interested in tracing their family histories, companies like Starr's have been working to give more granular results, and this is where things can get complicated.

So, how do these estimates work? Two major factors account for why they often vary between companies: reference populations and algorithms. Ethnicity estimates are by necessity all about patterns and probabilities; they are a process of comparing a person's DNA against that of known populations and looking for the best possible match. Reference populations are sets of people, typically culled from public datasets and from a company's own customers, who are meant to stand in for the world's historical populations anywhere from several hundred to a thousand years ago, because they have verifiably deep roots in those places. But the actual people chosen to represent those populations will vary from company to company, as will the number of them. The idea is to match the hundreds of thousands of DNA markers, or SNPs, in each customer's DNA up against these reference populations, looking for what Bennett Greenspan of FamilyTreeDNA calls a "best fit." But here the second factor comes into play: Each company has its own method of crunching the data. "Each company thinks, 'Our algorithm is the best,' but each algorithm has its pros and cons," says Yoav Naveh, who heads MyHeritage's DNA team.

Over time, companies have honed their algorithms and increased their reference populations with an eye toward making their results more accurate, and indeed, they're much better than they were a few years ago. In late 2018, for instance, AncestryDNA unveiled a new ethnicity prediction system. Previously, the company had been working with a reference population of three thousand people; now its reference population was more than sixteen thousand, augmented in great part by customers from its own databases whose deep roots in particular geographic areas had been robustly vetted and verified. In a flash, AncestryDNA's Benin and Togo reference population went from 60 to 224. In the old reference panel, there were 394 samples from the vaguely named "Asia East," but nothing was specifically marked as "Japan" or "China"; in the new panel were 592 Japanese samples and 620 Chinese ones, making it so that the company could better distinguish between these two groups of people.

The other thing that changed was the method by which the company looked at the DNA. Instead of looking at hundreds of thousands of individual markers along the chromosomes, Ancestry had begun splitting the genome into segments—1,001 short windows of between three and ten centimorgans in length—so that they could examine these markers in context.

"There is no French SNP or German SNP or Korean SNP," Starr told me in 2018. "Each SNP is just more or less likely to be from a different region. So when you look at individual bases"—that is, the As, Cs, Gs, and Ts that make up a person's genome sequence—"you're just looking at *most likely* ethnicity for that SNP. But when you look at a longer segment, you can smooth out the results." He went on: "Imagine you have a SNP and it's most likely Iberian—so, from Spain or Portugal. But when you look at it in the context of a longer stretch, what you see is that on each side, it's surrounded by SNPs from the UK, and you have this one Iberian one in the middle. What the algorithm can do now is correctly interpret the Iberian SNP"—it's actually only an *apparent* Iberian SNP—"as a UK SNP. Since the DNA of Europeans is all pretty similar, that Iberian one is probably found pretty often in the UK, as well."

Before AncestryDNA updated its approach, a consumer might have wound up with some small, mystifying percentage of Iberian ancestry, and wondered about family stories left untold. Now, when she looked at her results, the Iberian was simply gone, smoothed out by the new algorithm. Combined with another feature that looked for closer genetic ties within a larger ethnic group and combed users' family trees to identify common ancestral birthplaces and migration routes, the company was now promising to pinpoint more than 350 global regions and subregions for its customers. Since then, that number has grown even more.

Even so, the science was a matter of probability, which AncestryDNA tried to indicate by giving ranges to its estimated ethnicities. Just because the algorithm makes a prediction of a certain ethnicity "doesn't mean that that's the only answer," Starr said. "It's just the most likely answer." Likewise, 23andMe offers a "confidence slider," allowing customers to see what their estimates look like "at different probability cutoffs."

A few months after AncestryDNA's update, 23andMe updated its own results, telling customers it could now evaluate customers based on more than one thousand regions. Nothing in my DNA was any different, of course, but my results changed: The "Scandinavian" grew and the "French & German" fell. (If you want to get really technical, this apparent "French & German" ancestry likely did not, in my case, include any actual French ancestors; this was a kind of labeling imperfection, emerging from the fact that the two populations are geographically close and can be genetically similar.) That tiny bit of Korean that the company detected a year before had dropped down to 0.1 percent "South Asian" in a previous update and now disappeared altogether. So much for that lost Korean ancestor I'd been wondering about. Now, 23andMe informed me with astonishing specificity that the area in Sweden that the Scandinavian portion of my DNA looked most similar to was called Skåne County. I asked my dad and he told me that yes, this was where his grandmother had been born.

To attempt to read the past through the genes, you need more than knowledge of science, statistics, and algorithms. You need to understand history, and history is profoundly messy. MyHeritage's Naveh told me the knowledge of genealogists and population geneticists on staff is crucial because it is in the movement and mingling of populations that DNA remakes itself. Populations that in the past were geographically isolated (Naveh pointed to the Finns and the Japanese) or married among themselves (like Ashkenazi Jews) are relatively easy to pick out. But for other groups, it's harder. Wars, migrations, porous borders, shifting nation-states—all these things make it difficult to distinguish neighboring populations from one another, or to define them by modern standards. An ancestry prediction may say your people came from "Germany," but what we know as Germany only dates to

1871—your distant relative may have actually called Saxony or Prussia home. Different companies also define areas differently. Ancestry grouped England, Wales, and northwestern Europe into one category, Starr said, because "the English Channel isn't that big, and there was a lot of traffic between England and northwestern Europe, and so genetically they're more similar than you might have guessed." Perhaps in the future, with an even more robust reference panel from those parts of the world, the company's experts might be able to break down these populations even further.

Genetic genealogist T.L. Dixon has likened ethnicity estimates to cocktails, to be enjoyed responsibly. "My goal is to get beginner and intermediate genetic genealogists to think more like population geneticists ('5 percent of my DNA is similar to British Isles populations . . .') rather than genealogy astrologers ('I have 5 percent Irish in me . . .')," he writes on his blog, Roots & Recombinant DNA. Just because "a portion of your genes are similar to [those in] Ireland does not necessarily mean your ancestors are from Ireland," unless you can confirm it through the paper trail. It's not the customers' fault that they engage in such simplistic thinking, he adds: "It's the fault of the DNA companies, who falsely advertise that you can take a DNA test and it'll tell you what's in you."

CeCe Moore says she worries about the fuzziness of ancestry predictions. "I've had people contacting me and saying, 'My father must not be my father because I got British and he's German!'" In the absence of any other evidence, you can't make that leap based on an ethnicity estimate alone, she says. And she also worries that because ethnicity estimates get "a bad rap, and much of it deservedly, it has led to misconceptions about the reliability of [relative matching], which is solid science."

Another problem has to do with the "undersampling" of non-European populations, including those of African and Latin American ancestry, as well as of indigenous peoples. "No company actually has adequate African samples," says Dixon, who specializes in tracing African ancestry. Joanna Mountain, 23andMe's senior director of research, says undersampling goes back a century, to the early days of genetics research. "Most of the researchers have been in the US and Europe and they have studied people who have been in the clinics nearby, or in the military," Mountain says. "In early days, that tended to be people who were white." Razib Khan, a population geneticist and science writer who previously consulted for FamilyTreeDNA, told me he believes undersampling has persisted in the consumer genetics world in part because the majority of people who are seeking these tests are of European ancestry. "It's customer demand," he says. "It's a business decision."

In *MIT Technology Review*, journalist Antonio Regalado has written about how undersampling is a problem when scientists develop polygenic risk scores, which survey many thousands of locations across the genome to come up with a composite risk score for conditions like diabetes, some cancers, coronary artery disease, and irregular heartbeat, which are subject to complex genetic factors. "Because the algorithms were developed by using large databases of health and DNA data mostly from people of recent European ancestry, the tests aren't as accurate in other groups, like black Americans, Hispanics, and Asians," Regalado observes. Many scientists have been trying to rectify this undersampling, concerned that it will magnify existing health disparities. As bioethicist Thomas May pointed out to me, the current state of things is not only fundamentally unfair to the groups that are undersampled, though it is certainly that; it is also unfair to everyone. "If we don't have diversity in the people being tested, our data will be less rich and less useful," he says.

In the context of the ancestry results offered by a company like 23andMe, undersampling generally means that a person's estimates wind up being more regionally broad, Mountain says—instead of predicting that a person's ancestry is, say, Coastal West African, or Southern East African, their estimate might just say "Sub-Saharan Africa." "We did start thinking about rectifying the undersampling and lack of research on non-European populations pretty early in 23andMe's history," said Mountain, who heads up the company's research diversity efforts. Mountain's PhD in genetics focused on African populations, and as a Stanford professor, she studied genetics and linguistics among the peoples of Tanzania. She recalled a company phone call years ago between herself, Anne Wojcicki, and Henry Louis Gates Jr. during which the subject of 23andMe's genetic analysis for African Americans came up. She told them she was concerned about gaps in the company's service "because the research hasn't been done to support the product for African Americans. So, I'm not comfortable marketing as aggressively,'" Mountain recalled. "Anne said, 'Let's do the research that's necessary to fill in that gap.'"

That was the beginning of a series of initiatives to increase the diversity of the company's samples, both to increase its understanding of health issues among different populations and to improve its predictions about customers' origins. (The company's first big initiative, for instance, looked at the genetics of diseases that affect African Americans.) In some cases this has meant providing free kits to customers who can trace their ancestry to certain populations; in others, it has meant collaborating with researchers doing genetic work in parts of the world where 23andMe would like to improve its data and where research on such populations will improve the diversity of genetics research more generally. The bulk of its efforts have focused on Africa and Southeast and Central Asia.

Mountain told me that she worries about how consumers understand and interpret concepts like "race" and "ethnicity." She avoids using those words herself, concerned that they may invite reductionist thinking about how different we are from each other, and 23andMe itself favors "ancestry composition" when it returns results on where a consumer's roots may come from. Mountain says the company has at times consulted social scientists for advice on how to frame its work in order to avoid what's known as "racial reification"—that is, reifying, or reinforcing, the idea of essential racial differences. But, she added, the desire for nuance and scientific precision was sometimes in conflict with consumers' needs for results that were comprehensible and intuitive.

"They like the pie chart, we have found," Mountain says.

* * *

At a 2018 Toronto conference on the ethical, social, and privacy implications of genetic testing, a light-skinned blonde woman stood up and told the audience that she was black. Françoise Baylis is a bioethicist and philosopher at Dalhousie University in Canada. She told the audience that she knew some people questioned her identity when they looked at her; they might wonder whether she'd taken a DNA test to back up her claim. She had not. She did not need to. In the 1960s, her mother was "the first successful case of racial discrimination in employment in Canada. In testimony, my mother had to stand in court and answer the question 'Are you a Negro?'" If sometimes Baylis's own blackness was suspect in the eyes of others, she said, other times, it was her whiteness that was suspect. She was accused of "passing" when she was simply walking around in her own body. From this position of in-betweenness, and through her work looking at assisted reproductive technologies, she had thought deeply about questions of genetics and identity.

Baylis said the crux of her work was a theory that identity resides not in the body alone, nor in the mind alone. Rather, she said, she'd come to believe identity was a thing we work out in constant negotiation with others—"a narrative, relational account of who I am," a thing that shifts over time, with a person's experiences, and that is in part dependent on how other people respond to us. Identity is a lived thing, she said, a series of stories about a complex experience. She worried about beliefs that the age of genetic testing was giving rise to. "Identity is dynamic, it's social, it's cultural, it's political, it isn't this . . ." She gestured to the screen beside her, which showed a screengrab from the AncestryDNA website. "It isn't percentages," she said.

These issues around differing cultural, political, and scientific definitions of identity burst into the mainstream American political discussion less than a month later. Shortly before launching her presidential exploratory committee, Massachusetts senator Elizabeth Warren attempted to put to rest the taunts of President Donald Trump, who had mocked her claims of Cherokee and Delaware heritage with the nickname "Pocahontas" and, at one point, challenged her to prove it. Warren released the results of DNA analysis by Stanford University population geneticist Carlos Bustamante, showing "strong evidence" of Native American ancestry six to ten generations back. But instead of ending the matter, Warren's move succeeded in provoking raucous debate about who decides Native American identity, about the implications of politicizing claims of genetic ancestry, and about the conflation of culture and genetics.

Leaving aside the other big criticism of Warren's decision—that she "caved" to the race-baiting of the president—her decision angered many Native Americans. The experiences of the Havasupai with University of Arizona researchers, as well as other experiences in which Native and indigenous peoples have felt exploited by scientists, or have seen their cultures appropriated by white people, seemed to distill into this one moment. In the 1800s, naturalist Samuel Morton, whom the journalist Elizabeth Kolbert has called "the father of scientific racism," collected and studied skulls in order to divide humanity into a hierarchy of five races, with Native Americans second to last and blacks (or "Ethiopians") at the very bottom. The University of Alberta's Kim TallBear, who researches ideas of race in genetic science, draws a straight line from history like that to "old-school notions of racial purity that continue to inhabit genetics even today." Indeed, she says, Native Americans are understudied by geneticists in part because of tribes' resistance to this research, based on a history of mistrust.

The Cherokee Nation secretary of state, Chuck Hoskin Jr., issued a statement that "Senator Warren is undermining tribal interests with her continued claims of tribal heritage," and did interviews explaining that his tribe had its own way of determining membership, primarily through a "rigorous process" of documentation tying a person to the

Dawes Rolls, "a turn-of-the-twentieth-century census of Cherokees living in what was to become northeast Oklahoma."

"We fought long and hard for this status," Hoskin told NPR. "It's a legal status that we use to fight for the rights of Native children, to protect our lands. And when someone comes in and boasts that they took a DNA test so they're Indian, we think that erodes what it means to be a Native American in this country, even if they don't claim a specific tribal nation citizenship."

Do we define ethnicity by way of cultural heritage or genetic assessment? And who gets to decide? TallBear, a member of Sisseton Wahpeton Oyate, a South Dakota tribe, says suggesting that genetics alone can define Native American identity is in direct conflict not only with Native Americans' political authority but also with how Native Americans themselves define kinship—as a web of relationships tied to living people and to ancestors kept alive through stories. She says that DNA-only definitions are shallow and that autosomal DNA testing may even exclude some legitimate tribal members, given the fact that we lose some of the genetic material of our ancestors with each passing generation. She has written that Warren "and much of the US American public privilege the voices of (mostly white) genome scientists and implicitly cede to them the power to define Indigenous identity." (Warren later apologized to the principal chief of the Cherokee Nation and told the *Washington Post* she regretted claiming Native American heritage.)

"The question is not whether her DNA analysis is accurate," Columbia University sociologist Alondra Nelson wrote in the wake of Warren's release of her DNA results. "It's whether it can tell us anything meaningful about identity. The truth is that sets of DNA markers cannot tell us who we really are because genetic data is technical and identity is social." The concern voiced by Nelson, TallBear, and others is that believing too fervently in the truth-telling ability of the genes, as interpreted by imperfect science and imperfect human beings, may flatten and oversimplify our understanding of who we are.

And yet, as Joanna Mountain observed, and as much of the advertising around genetic ancestry testing testifies, these messages are also appealing. Roberta Estes told me she is troubled by that AncestryDNA commercial in which a guy named Kyle trades in his lederhosen for a kilt, both because it ignores the limitations of the current science of biogeographical ancestry analysis and because it offers a facile answer to a complex and deeply personal question.

"Believe me, we all hate it," Estes says. "The reason Ancestry runs it is because it's so effective for sales. Because people are like, 'I can take this test and I can get The Answer.' . . . Well, you know what? There is no The Answer." When it comes to the question of *Who am I?*, "there are many answers and a lot of it depends on how you ask the question."

10
WHAT TO CLAIM

How does consumer ancestry testing affect how we think about our genetic differences? Joanna Mountain of 23andMe says she's long wondered about that. Her company didn't set out to help people explore their ethnic and racial identities, she says, but over time it became clear that this was exactly the work they were doing. In 2008, the company debuted something called Ancestry Painting, which attempted to show customers' ancestral mixes by dividing their chromosomes into crude geographic categories—Africa, Asia, and Europe—based on the technology available at the time. "It was very disturbing for me because it aligned closely with the big three racial categories," Mountain said. She worried such a simplistic breakdown might reinforce people's notions of essential biological difference along racial lines.

But in time, the company's results improved, becoming more refined, and Mountain started hearing more and more stories about consumers discovering unexpected ancestral stories, including ones that crossed the color line. And Mountain wondered what effect such results might be having. When consumers discovered this diversity within themselves, would that make them more flexible in how they thought about race? Meanwhile, 23andMe was studying this diversity, as well as the legacy of historical constructions of race. In 2015, 23andMe population geneticist Kasia Bryc, along with Mountain, David Reich, and others, published a study from the company's consumer data revealing that close to 4 percent of those who identified as white Americans had at least 1 percent African ancestry, consistent with an African ancestor within the last eleven generations or so. (In southern states like South Carolina and Louisiana, about 12 percent of whites had 1 percent African ancestry or more.) Henry Louis Gates Jr. observed of these findings: "What hasn't been confirmed until now is how many self-identified 'white' women and men are walking around today with recent 'hidden' African ancestry in their families—you know, white people, who, at least according to the old, notorious 'one-drop rule' of the Jim Crow era, would have been considered legally 'black.'"

Research is demonstrating the importance of how we frame discussions of genetic difference. In *DNA Is Not Destiny*, cultural psychologist Steven J. Heine writes about his work showing that American subjects who tend to believe that genes determine life outcomes also tend to be more bigoted toward African Americans. "Thinking about genes as underlying essences and having racist thoughts seem to go hand in hand," he writes. But if the way we approach such questions can doom us, it can also save us. Heine points to more research showing that when we focus on our sameness, we do better, as in his study showing that adults prompted to consider genetic commonalities are more likely to offer environmental explanations for racial stereotypes.

Mountain hasn't yet been able to answer the question of what impact companies like hers are having on attitudes about genetic difference. She helped design a study conducted by Northwestern University that found a third of Americans believed that genetics "totally" determines racial identity, and she hopes future collaborations can gauge whether DNA testing causes this belief to grow or shrink. Beyond this, research on the topic is in the early stages. A 2014 study out of Columbia University found that having subjects read a newspaper article about genetic ancestry tests appeared to "magnify the degree, generality, profundity, and essentialness of the racial differences people perceive to exist," but additional work suggests that the effects are not so clear-cut. Professor Anita Foeman of Pennsylvania's West Chester University, who studies people's experiences of ancestry testing, told me she thinks the technology is opening up much more complex and sophisticated conversations about race. In one talk I watched, she put up pictures of a diverse range of people and invited the audience to speculate on who matched a particular pie chart. The blonde woman was part Asian, according to testing, while the child of Indian immigrants had European ancestry. It was impossible to predict, which was exactly Foeman's point—the genomics age can't help but blow up our neat categories.

Sociologist Wendy Roth, now at the University of Pennsylvania, has found evidence for both theories: that ancestry testing reinforces our sense of immutable racial categories, and that it breaks them down. The difference depends on the test-takers themselves. In a forthcoming paper, she found that a tester's knowledge of genetics made the biggest difference in how he or she understood the implications of a test. Among more than eight hundred subjects, Roth found that for those with high knowledge of genetics, taking a consumer DNA test generally reduced their belief in genetic essentialism. For those who went in with very low knowledge of genetics, "there's some indication that it may actually increase belief in genetic essentialism," she says. In previous research, Roth had observed that people were also motivated toward racial essentialism—or away from it—based on what they wanted to believe. One subject pleased to discover apparent Basque origins said it reinforced her sense that race was something revealed by DNA, while another subject, who wasn't eager to embrace the Middle Eastern origins suggested by his test, preferred to see race as a social and economic construct.

"The way people interpret scientific evidence is not neutral," Roth told one audience, likening the way people interpret ancestry tests to what they see when they look at Rorschach tests. Which gets us to another point: How people determine what portions of their pie charts to claim and what portions to ignore is a complex process indeed.

* * *

People filter their ethnicity estimates through a complex web of cultural and personal realities. They bring ideas about truth and authenticity, fantasies about different cultures, and notions of the past. They bring their loyalties and longings and resentments, not to mention their genealogical knowledge for times when the paper trail clashes with the genetic one. In other words, they don't accept their results uncritically.

But sometimes, those pie charts can offer them just what they've been looking for. In 23andMe's podcast, *Spit*, company co-founder Anne Wojcicki tells the story of an adoptee who wrote to say how meaningful the company's ancestry composition report had been in her life. Upon discovering her true genetic heritage, she had quit her Bay Area job and gone to live with her people, an indigenous population near the Arctic Circle. She told Wojcicki she felt at home for the first time. "And I take this story," Wojcicki recounted, "and I walk over to my research team and I go, '*You better be right.*'"

Indeed, they'd better. Because consumer genomics is changing how some of us see ourselves. Wendy Roth has found that most testers tend to stick with the ethnic identity they had before the test, even when results reveal something unexpected. But for the sizable minority motivated to consider another identity, they pick and choose what to claim, often weaving the new knowledge in with an existing identity. What they choose is based on their aspirations and on how others respond to their claims to a newfound ancestry. In that sense, their ethnic identities are indeed—as philosopher Françoise Baylis describes it—a kind of ongoing social negotiation.

Interviewing one hundred early adopters of DNA testing, Roth and colleagues found that white testers seemed more driven than other groups to seek out new ethnic and racial affiliations. These seekers offered a host of explanations. They spoke of a desire to gain a sense of belonging and identity, and to distinguish themselves from what they saw as the blandness of being "just white." Roth told me she was struck in her interviews by how many European Americans had lost connection to their ancestral roots.

But Roth also noticed that white testers were drawn to the "cultural cachet" of being multiracial. "For our White respondents, it is not being non-White that is most desirable, but having a bit of something non-White," Roth observed. One subject, an adoptee who initially thought of herself as white, took a mitochondrial DNA test that suggested an African haplogroup. She incorporated that into her identity and began telling those close to her that she was "mixed," even as she allowed that she did not have to face the social consequences of blackness "because I don't have Black skin."

Roth found that testers who identified as black or African American were far less inclined to incorporate new ancestral knowledge into their identities. In part, that's because they tended to identify strongly and positively with their existing identities; unlike white respondents, they did not describe their race as boring and plain. As well, Roth says, the routine rape of black women by white men under slavery, and a historical construction of race based on notions of white purity, meant the African Americans she interviewed had always assumed they were more than one thing. Finding evidence of other ancestries "doesn't change the label that society applies to them and it doesn't change the way that they think about what it means to be African American," Roth says.

In her 2015 paper, 23andMe's Kasia Bryc found that African Americans in her study tended to have about 73 percent African ancestry and the rest mainly European. Bryc's study found that more than half of the contribution of European ancestry to African Americans occurred before 1860—in other words, during the time of slavery. History is ugly, Anita Foeman says, and sometimes, the ugliest actions of our ancestors are what carried our genes into the future—leaving African Americans who discover their great-great-something owned slaves and committed rape to grapple with the fact that "genetically, you're related to those people. It's not *they*; it's *you*," Foeman says. "That messes you up." This ugly reality is why so many people were outraged by a 2019 AncestryDNA ad that they argued whitewashed the systematic sexual violence, coercion, and exploitation under slavery, offering a romanticized explanation for why many African Americans have white ancestry. "Abigail, we can escape to the North," an 1800s-era white man tells a black woman in the ad, holding out a wedding ring. "There is a place we can be together, across the border." For black Americans, "DNA and documentary ancestry information is as painful and traumatic as it is illuminating," journalist Kimberly Atkins pointed out in an incredulous tweet. "These are not love stories." Ancestry apologized and said it was pulling the ad.

The DNA age forces reconciliation with the past, and with the countless ways in which it lingers and shapes our perspectives now. It forces reconciliation with what our culture, politics, and policies have told us about what race means. One of Roth's African American subjects described a sociopolitical motivation for not embracing the apparent European ancestry revealed by a DNA test: "African Americans number only about maybe, what, 36 or 37 million? . . . I don't want to dilute or make that count fall down, because my whole experience has been African American." Besides, he said, he did not want to run away from the historical traumas of his people. A subject in a different study conducted by Foeman put it another way: "My momma used to tell me, 'If the cops stop you, they're not going to ask if your momma is from Ireland,'" said a dreadlocked university professor who identifies as black, after getting DNA results suggesting he was primarily of European descent. "Even though I know that race is a social construction, it is as real as oxygen," he said.

Like Roth, Foeman has spent years looking at race through the lens of genetic testing. She's found that even as people of color expect more diversity in their ancestral backgrounds, African American testers routinely underestimate just how much European ancestry they have. Foeman, who founded West Chester University's DNA Discussion Project, which since 2006 has tested more than three thousand people, has found women are generally more flexible about their racial identities, as are younger people compared to older ones. She says white and black subjects both tend to overestimate how much Native American ancestry they'll find—a belief she and colleagues believe is partly connected to a romanticization of what it means to be Native American and to a desire for an "indisputable" claim on this country. What we want to be becomes what we think we are. Identity is aspirational.

On the one hand, Wendy Roth finds her work demonstrating people's selectivity heartening. It suggests that scientific knowledge is just part of how people construct their identities. On the other hand, who picks and chooses what new information to incorporate into their identity is telling. The ability of white people to claim multiraciality or blackness as something symbolic and "exotic," to be revealed on a case-by-case basis when it suits them, means genetic ancestry testing may be reinforcing the privileges of those who already experience them.

It's clear that law, culture, and practice have not caught up to an era in which we can easily and cheaply test for genetic heritage. There have been a few high-profile cases of people using genetic testing to claim benefits or other advantages associated with minority status, although these may be outliers; Roth says she did not encounter this as an incentive for testing among her subjects. An investigation by the Canadian Broadcasting Corporation found some Canadians were claiming indigenous status—possibly as a means to getting tax exemptions—after sending their cheek swabs to a company that appeared to reliably return reports of sizable Native American DNA to everyone. (The company even declared one man's chihuahua to be 12 percent Abenaki and 8 percent Mohawk.) And a Washington State business owner named Ralph Taylor applied to have his company certified through the federal Disadvantaged Business Enterprise program on the strength of a DNA test; in 2018, after Taylor was approved as a minority for a state program and denied by the federal program, he sued. The case opened up questions about the reliability of ancestry estimates, how the government determines minority status, and whether someone like Taylor—who, according to the *Seattle Times*, "lived most of his life as a white man," but claims he is 4 percent sub-Saharan African and 6 percent indigenous American—should have to prove that he's suffered discrimination. Sociologist Alondra Nelson points out that seeing black or white identity as embedded in genetic markers alone misses the point that race is a social construction. "You have two facets of identity: who you think you are and what other people say you are," she told the *Seattle Times*. "Generations have been disadvantaged based on what they look like, how they talk, or where they come from."

And what happens when a person is *not* motivated to claim something suggested by an ancestry test? Foeman thinks of our family history narratives as akin to lullabies, with the power to lull us into a sense "of security and predictability"; DNA tests can disrupt these stories. A woman in Foeman's project who hailed from a family proud of its Italian heritage reported that her father hung up when she informed him of test results saying she was only 19 percent Italian. Another woman, who identifies as Iraqi, was "outraged" by the results of a test telling her she was largely European.

In an extreme example, members of the racist far-right have used this consumer technology to attempt to prove their beliefs about racial purity—ones that nineteenth-century naturalist Samuel Morton, with his pseudoscientific ranking of skulls, would have recognized. Sociologists Aaron Panofsky and Joan Donovan decided to study what happens when genetic testing proposes a profound cognitive disconnect: when you learn you *are* the people you hate. Analyzing years of posts on the white nationalist website Stormfront, Panofsky and Donovan found that, in some cases, posters who confessed to something other than wholly European ancestry were shunned and shamed. But far more often, the community offered "emotionally supportive" responses (as in "I wouldn't worry about it. When you look in the mirror, do you see a Jew?") or attempted to reinterpret a tester's results in a better light. Sometimes they pointed to the very real imperfections of ancestry estimates as proof that unwanted findings were "statistical error," or theorized that the genetic testing industry is a Jewish-controlled conspiracy intended to confuse whites and spread multiculturalism.

In other words, this is an old story; selectivity is not a feature of the genomics age alone. As historian François Weil points out, genealogy has long been about not just who we think we are but who we want to be, going back to the days when Americans purchased suspect pedigrees in hopes of borrowing a little aristocratic luster. In *She Has Her Mother's Laugh*, his book on the science and culture of heredity, science journalist Carl Zimmer writes of an early-1800s Massachusetts genealogist who, in drawing her family tree, simply left off those branches descended from a couple considered "near idiocy." I've noticed people often explain surprising ancestry findings by looking for confirmation in preexisting preferences—as if they somehow always sensed their own essential Korean-ness or Italian-ness. This may have to do with the way human beings look for patterns to make sense of our lives, and the fact that—as professional fortune-tellers know—we're willing to do a lot of interpretation and filling-in-of-the-blanks when we're motivated to believe something. "Now I know why I like tulips so much, why I love those Belgian chocolates, and why I eat so many French fries," writes one man upon discovering Western European ancestry. "Now I know why I have always liked the Beatles!" writes one woman about her apparent British ancestry.

But if such explanations can seem facile, DNA testing can also be the means by which profound understandings of identity are forged. Many seekers make nuanced decisions about how much of their genetic backgrounds to own. They wrestle with what those pie charts mean. Is it ever too late to identify with one's roots, if you grew up as something else? And how do shifting loves and loyalties influence this? One woman I'll call Nancy told me that when she finally reconnected with her father in adulthood, and he told her how he'd tried to remain in her life as her mother refused him contact and moved from town to town, she also found herself motivated to claim the African ancestry he'd passed down to her. Nancy told me her upbringing as a white girl had left her feeling confused about being part black—"having this identity but not being able to truly embrace it." She straightens her curly hair; people ask if she's Italian or Hispanic. But after reconnecting with her father, she went to a doctor's appointment and for the first time checked white *and* black on the form. She did this out of respect for her father, she told me. If he wanted to know her, she wanted to claim him.

We may be selective in how we identify in the wake of genetic testing, but this doesn't mean we're wrong to be selective. We are the products of culture, memory, family. DNA is part of the picture, but it isn't everything. As Roberta Estes puts it, "It's not love."

* * *

T. L. Dixon wanted his own *Roots* story. He wanted to know his people.

Dixon was adopted, and traditional genealogy had helped him track down his biological sister in adulthood and to learn that his birth mother had passed away. He wanted to know more about his genetic family and its past, but further back the paper trail became difficult to follow. Dixon did not know his ancestors' names, did not have records of them the way so many other Americans do. He did not know when they were taken from their country, or even what their country was. He did not know what ship they were put on, what plantation they were taken to, what children they bore, how their lives began and how they ended. The profound losses of slavery were echoed down to Dixon in yet one more loss: the paucity of genealogical records about his enslaved ancestors. As the science journalist Christine Kenneally points out in her wide-ranging book, *The Invisible History of the Human Race*, "the thing about records is that they tend to proliferate as a matter of course around people with power." Those without power are silenced by the lack of records.

Dixon, a social worker who was previously a journalist, has written on his blog, *Roots & Recombinant DNA*, about how the 2018 movie *Black Panther* sparked a kind of *Roots* 2.0 for African Americans who want to find their own "Wakanda"—the fictional African country that is home to the superhero. And yet, where would they start? How could they find out more? "For the majority of African descendants in the Americas, our 'Africa' might as well be in Wakanda because our actual ethnic origins felt like a sort of mythical—and stereotypical—place," he observes.

The difficulties of African Americans trying to trace their ancestries are manifold. Many hit a brick wall trying to trace their ancestors before 1870 because that federal census, the first done after the Civil War, was the first to name former slaves. Before that, FamilySearch's spokesman Paul Nauta has said, enslaved persons "showed up as ticks or hash marks on paper." And another legacy of being treated as chattel means that people in a person's family tree may be spread out all over the country, the result of "enslaved Africans being moved around from place to place," Dixon says.

The longing, and the limitations of genealogical records, are what made Dixon turn to saliva back in 2012, and what made him a genetic genealogist. Through the bigger databases, he was able to find and make contact with several distant cousins who are of recent African descent. The genetic heritage they have in common with Dixon traces back to a time before his ancestors were imprisoned and shipped abroad, and they provide a kind of window into what their lives might have been like before the profound violence and disruption of the transatlantic slave trade.

On his blog, Dixon documents how he used these African cousins to trace his likely ancestry not just to broad areas within the continent but to particular regions of countries, including Cameroon, Guinea, Nigeria, and Ghana—and to specific ethnic groups within those regions. On 23andMe, for instance, Dixon was able to see he shared twelve centimorgans of genetic material with one distant cousin along their seventh chromosomes. He confirmed and contextualized that little piece of DNA by triangulating it with his American relatives, and reached out to his African match, who told Dixon about his people, the Fulani, and gave Dixon family surnames. Drawing on books, online articles, and old newspaper clippings, Dixon was able to understand the historical forces that likely shaped the life of his Fulani ancestor.

Dixon was still trying to find more about his specific ancestors, he told me when we spoke during the summer of 2018. He'd recently heard about a new collection of slave records he hoped to consult. But what he'd been able to achieve was very meaningful to him—and would not have been possible without DNA. "It was actually very fulfilling and amazing that I was able to find my actual African roots and not just say, 'Oh, I'm African,'" he said.

In her book *The Social Life of DNA*, Columbia University sociologist Alondra Nelson writes about how African American early adopters of genetic testing have used this technology to access their own histories. Nelson was struck by this, given a history that has left many African Americans mistrustful of medical research: eugenics and forced sterilization; scientific exploitation like the horrifying Tuskegee syphilis experiment, conducted by the US Public Health Service for decades on

African American men; a long history of segregated health care. Despite "black communities' accumulated mistrust of science," despite awareness about how intertwined the history of genetics is with scientific racism, these early adopters saw something special in the opportunity posed by DNA testing. Nelson argues that these early adopters were engaged in their own small-scale journeys toward reconciliation with the pain of the past, "in which DNA test results are used as a form of testimony."

Recreational DNA testing can be an unflinching history lesson. Marissa Kinsey, a graduate student in library sciences in Queens, told me about wrestling with the implications of this history up close after an undergraduate assignment to create a family tree sparked an interest in genealogy and DNA. (Kinsey is transgender and asked me to use gender-neutral pronouns.) Through genetic genealogy, Kinsey was able to confirm the outlines of their family's oral history: An enslaved great-great-great-grandmother named Isariah bore a baby girl named Fannie by her master. From probate documents, Kinsey was able to deduce that Isariah was likely about fifteen when she had Fannie in the early 1850s, and not long after that, mother and child were separated. The little girl, valued at one hundred dollars, was placed in "Lot 1" and parceled out to the white slave owner who was very likely her own father. The mother, valued at five hundred dollars, was in the next lot and went to a different family member. Kinsey was able to figure out the likely identity of this slavemaster and to connect with a white cousin descended from this same man. Kinsey told me of spending time in an archive in North Carolina with their mother, sorting through wills and slaves' bills of sale, the two of them by turns distraught, angry, and driven, wondering about the lives of these enslaved ancestors. The second time I spoke with Kinsey, almost a year after we first spoke, they were in the process of legally changing their name to Riah Lee Kinsey, both to reflect a gender-neutral identity and to honor the ancestor who had so much taken from her.

Kinsey told me they struggled with how to view the white slave owner they descended from. "What kind of man could sell a woman away from her child, but then take that child in and raise her?" Sometimes, Kinsey thought that perhaps "there was a crumb of human kindness in him that he actually was able to raise this child." And then, Kinsey wondered about this desire to find something good in him—about the possibility that perhaps they were driven to look for this good in this man because they were descended from him, about the fact that some things are just too terrible to look upon, and so we refuse to see the whole picture.

The rise of genealogy and DNA testing may be making it increasingly difficult for white Americans to engage in historical amnesia, to fantasize about their own family histories without ever considering how slavery might have figured into them. In 2012, when the *New York Times* reported that new research showed that Michelle Obama was descended from a slave owner's son and one of the family's slaves, descendants of the white line had varying reactions to whether they'd like to meet their extended black family or, indeed, be quoted in the newspaper about the news. One struggled with how to understand the relationship between her great-grandfather and Obama's great-great-great-grandmother. "The idea that one of our ancestors raped a slave . . ." she said, trailing off. Kinsey told me they have spoken with their white cousin, to whom they are related through that ancestral slavemaster. "I feel like it's a little bit more awkward for her because she hadn't known about us," Kinsey told me. "We kind of just appeared out of nowhere, and I'm pretty sure that's not what she was banking on."

* * *

Rosario Castronovo is still seeking. Over the years, the would-be Sicilian—who is, he knows now, the son of a biracial woman who hid her ethnic identity—has consulted spiritual mediums and woken in the night believing he'd heard his grandfather's voice. The last time I talked to him, he was planning to change his name again, this time to Jerome Lafayette Naramore, to honor the ancestral discoveries he'd made through genetic genealogy. "Some people say what's in a name?"

says Rosario, a restaurant server and union officer who lives in Manhattan. "I think everything is in a name."

Rosario struggles with how to think about himself and how to present himself to the world. "It's kind of weird for a guy who presents as white to say, 'I'm black,'" Rosario said. "I'm still getting used to it." Could he even claim blackness, he wondered, without that lived experience and without any of the implications of what it means to be perceived as a black man in contemporary America? "I'll never know what it's like to be pulled over by a police officer and fear for my life," he said. "Can I call myself black and never have to experience it? It almost seems unfair, or like I'm pandering." Yet his mother's elderly cousin, who is more fair-skinned than he—and, according to her DNA results, has less African ancestry than his mother—is quite clear in her identity as a black woman. "It's a social construct," Rosario told me. "I raised myself as an Italian man, I immersed myself in Italian culture. What do I do now?"

We were talking on Independence Day, and Rosario told me the more history he learned about his black family, about the lives of black people in Vermont and in the rest of America, the more he wondered what he should be celebrating on this day, and if he should be celebrating at all. "I have learned more about African American history in this country, and the more granular you get, the more bitter you become," he said. "I never got to know my people."

And yet genetic genealogy had also given him an opportunity to know about what had been hidden, to wrestle with his mother's pain, with his grandparents' sacrifices, and with the wrongs done to his family. DNA testing had given back to him and to his mother a little of what was stolen by the past.

Rosario's mother has Alzheimer's now, but there are moments of lucidity. One day, Rosario took her to a quiet spot in a church not far from her assisted living home, and handed her a framed copy of the only photograph he had of her father: the man's mugshot, from when he had served time for the crime of adultery with a white woman—for the crime, really, of creating Rosario's mother.

The old woman sat in a pew and stared at the image. She wiped her eyes and held the photograph to her chest and kissed it. She couldn't get over how handsome her father was.

"This is so wonderful," she said. "I could just sit here and cry."

Rosario sat down in the pew and told his mother they could do that together.

11

THE MYSTERY OF JIM COLLINS

On September 25, 2012, Alice went to one of FamilySearch's family history centers, the satellite offices offering access to the LDS Church's treasure trove of genealogical materials, not far from her house in Vancouver, Washington. She was doing research on her paternal grandparents' last names, investigating her theory that one or both of her grandparents were Jews from eastern Europe who had somehow come to the United States via Ireland, Irish-izing their names along the way. It seemed a thin tendril, but Alice's policy was that any plausible possibility had to be explored.

So anyway, there she was, trying to find a record of her paternal grandfather's birth in Ireland, when she got an email on her phone saying her paternal first cousin Pete's DNA results had come in. She got out of there real fast and raced home. She logged on, navigated to Pete's account, which she was administering, and looked at his results.

And she discovered that Pete was not related to her. Not one tiny little bit.

Alice felt crushed. But they *were* cousins! She was Younger and he was Elder, separated by precisely fifteen years. Younger was so fond of Elder. She'd lived most of her life knowing she and Pete shared a birthday, and not knowing what had happened to him. "I wonder what happened to Pete," her father had said on his last day on earth. And then Pete and Alice had found each other late in life by way of crossed letters that seemed like kismet. Alice could not bear to lose the cousin she'd only recently found. She was, in her own word, "devastated. And I was afraid he was going to reject me because we were no longer biological cousins."

Now, Younger called Elder, and told him the news. She was relieved by his response. "He was sad, but he also told me I was the best cousin he ever had."

Alice talked to Gerry. They knew what this meant. The sisters drafted an email to their brothers and titled it "Important Family Information."

"Most of you know I took a DNA test a few months ago," Alice wrote. She walked them through everything that had happened since: the unexpected ethnicity results from AncestryDNA; her fear that she wasn't Jim's daughter; the 23andMe results, which eased that fear by proving they had a different mystery on their hands; the eureka moment along Bill's X chromosome; and now, the latest confounding finding. In typical Alice style, it was thick with facts and details, but the bottom line was this: "We share 0.0 percent of our genes with Pete!" she wrote. "Furthermore, his admixture indicates that his origins are entirely in the British Isles, exactly what I expected to find for us. Gerry and I interpret this to mean that Dad is not the biological son of Katie Kennedy and John Josef Collins, but our aunt Kitty *was*."

In other words, Jim Collins had a different set of parents than his beloved sister. Looking back, the sisters could see there were always puzzles. Gazing at old family photos on her wall, Gerry had long believed she could see bits of herself and her siblings in the faces on her mother's side of the family. There was the Nisbet nose up through the generations, a steady marker of Nisbet genes. But on her father's side, she could not see that same genetic legacy. In the photo she had of her paternal grandfather, John Josef, he had deep-set eyes, a long face, sharp cheekbones, and a cleft chin—he looked like no one in her immediate family. Plus, John Josef was said to have been tall. Gerry had gone so far as to blow up this picture of her purported grandfather, alongside a photo of her father, and take them to Thursday morning bowling. "Can you believe they are supposed to be father and son?" she'd asked her friends.

But photographs can be poor proxies; comparing the living was rather more dramatic. Gerry remembers meeting their uncle John, her father's brother, when he came to visit California, in 1975, and noticing how different he was from her father. Uncle John was about six foot one, and his face and build were entirely different, with the long, narrow face that John Josef had and that she'd recognize decades later in her first cousin Pete. Her father, Jim, was eight inches shorter, with a round face, dimples, and an entirely different build. Even the brothers' mannerisms seemed different, Gerry thought. Even so, it took a family outsider to point out the obvious.

"I don't understand how your dad can be that short," Roger Wiggins, Gerry's now-husband, told his fiancée.

"Well, my dad was in the orphanage," Gerry said, parroting the family line. "He was malnourished."

Even then, though, no one suspected that Jim and John might not be related. It was the sort of thing that seemed obvious only in retrospect, because sometimes, you don't see something "unless you're really looking for it," Jim Collins's son Jim the younger told me. "There are always people in families that look a little bit different."

After Alice and Gerry sent their family-wide email, what ensued was a lengthy exchange of emails among the seven siblings, spread out across the country from Washington State to California, Minnesota, Colorado, and North Carolina. They talked about what to do now, and they talked about the past and about what this all meant. They thanked Alice for her doggedness. They talked about the possibility that they might have biological family they'd never known—a boon for the children of a man who'd spent his childhood in an orphanage and didn't have much in the way of extended family. They talked about their identity as Irish Catholics, what had changed and what hadn't. They talked about how they were lucky to be a family, however it had happened. "Does this mean we get more holidays to celebrate?" one brother wrote, before expounding on the elusive nature of truth.

Alice wrote that she never could have imagined this possibility when she first sent away for a DNA test. It was a strange time, she would say later. "When I realized that my dad was not who he thought he was, I really lost all my identity," she said. "I felt adrift. I didn't know who I was—you know, who I really was."

It had been only a few months since the first DNA test, and it would be years before Alice and her siblings would finally unravel the truth about their father and themselves. But even so, they'd learned so much more in those few months than they ever could have anticipated—so much about science and family and the vagaries of life, and so much about the profound and disruptive power of DNA testing.

* * *

For months, Alice had been lurking on the Yahoo group DNA-NEWBIE, reading about the travails and techniques of other seekers. She knew she needed help now, so she posted for the first time. Her journey had begun on a lark, she wrote, and was now "extremely personal." She explained how she'd been certain the original AncestryDNA test was wrong, how she'd taken the 23andMe test in hopes of disproving it, and how she was now testing a brother's Y-DNA to try to find her biological grandfather. Was she doing the right things? Were there other tests she could be pursuing? Could anyone help?

CeCe Moore reached out to Alice, and they began to email and exchange information. The genetic genealogist had helped many adoptees trace their way to their biological parents, and she saw Alice's situation as similar. The late Jim Collins was in the position of many adoptees, with his biological parents unknown. Even back in 2012, when the databases of people who'd undergone genetic testing were much smaller than they are now, some adoptees were having luck by searching among their genetic cousins, using segment triangulation and family trees to deduce the identity of their biological parents.

But looking at some of Alice's autosomal DNA matches, which Alice gave her access to, Moore could see this was going to be a challenge. Ashkenazi Jews are considered an endogamous population, which means people marry within the group and, sometimes, within families, with cousins marrying cousins. There are a number of populations for whom this is an issue, including Cajuns, Polynesians, the Amish, French Canadians, and people descended from early colonial American populations, posing challenges to those among them who wish to use genetics in their genealogical work. Endogamy inflates the relationship predictions for Ashkenazi Jews, leading to the appearance of many seemingly close cousins who are, in fact, more distant cousins related to a tester in more than one way. And Alice's genetic match list along her father's side "was just a jumble of Ashkenazi Jewish endogamous spaghetti," Moore told me.

One by one, all of Alice's siblings were submitting their saliva to 23andMe, in hopes that having more Collins DNA and more genetic cousins to look at might help their search. As each one got results, Alice would upload the autosomal DNA data to FamilyTreeDNA, which accepts transfers from other companies, in hopes of maximizing the pool of relatives they might match to.

She signed up to take a class outside Seattle. Billed as a class for helping adoptees, it refined techniques Alice was already starting to use, which she'd learned about on DNA-NEWBIE and elsewhere. The teacher, a search angel named Diane Harman-Hoog, had been helping adoptees for free since before autosomal DNA testing was an option, and eventually became involved with co-founding the volunteer website DNAAdoption.org. She had previously retired from her work as a manager for Microsoft, and had a background in mathematics and an MBA in information technology and finance. Like Alice, Diane Harman-Hoog had spent years as the only woman in the room. And like Alice, she was not afraid of data.

Diane's class, held for two days at a local library in early 2013, boasted a table of contents that would have intimidated many, with such items as "Odds of Segment Persistence," "X Chromosome Inheritance Charts," and "Phylogenetic Trees." The heart of the class was a methodology for working with the FamilyTreeDNA database that Diane and others had only recently researched and formalized.

Diane sent me an early version of "The Methodology," as it is known, and I found the nineteen pages of instructions absolutely overwhelming. Suffice to say, a seeker would have to be pretty motivated to use it, particularly if she weren't tech-savvy. Adoptees using autosomal DNA back in the "early days"—which were, astonishingly, less than ten years ago—really had to work for their discoveries, which were based off distant cousin matches detected by shared genetic segments in chromosome browsers. To start with, the seeker needed to input and organize a ton of data on spreadsheets, which involved "several tedious tasks," Diane explained in her paper. But if you could complete the grunt work, the beauty of the spreadsheets was the ability to organize. The little groupings of relatives with shared genetic segments were placed together, along with their names, the length of the shared segment, and the locations of those segments along a particular chromosome. Then, starting with relatives with the longest overlapping segments—in other words, those likely to be closer relatives—the seeker reached out to ask for a family tree or any other information. This made the process a collaborative one, dependent on the time, interest, and accessibility of other family researchers in those early databases. Then, the seeker used genealogical research to expand on her cousins' trees, going back in time to look for common ancestors—perhaps as far back as 1600 if she was trying to understand her relationship to a fifth cousin. "Eventually," Diane wrote, "you will want to try to piece together as many of these as you can into one tree." With enough relatives and ancestral lines, the hope was to find an ancestor to the adoptee, and use that to determine at least one side of the adoptee's birth family.

The challenge of obtaining and synthesizing all this data was immense. And too often, Alice was learning, cousins wouldn't respond to requests to compare genomes or share family information. The trees of genetic cousins, when they were available, weren't always in great shape; Diane told me she sometimes worked with information written "on torn scraps of paper and paper plates." And then there was the problem of endogamy. At the top of Diane's "Methodology" was a caution that Alice read but which did not deter her, because Diane's approach was literally the only option for hacking her DNA. "Unfortunately," it said, this approach "probably will not work with Ashkenazi lines."

* * *

Alice and her siblings had eliminated a lot of theories by this point—from the most likely ones to the far-fetched. If the mystery of

Jim's ancestry didn't harken back to Europe, or to a New World affair, they needed to look to when Jim was young. His early years had been marked by turbulence and upheaval. It seemed possible something had happened then.

In the fall of 2012, Alice and her siblings decided to write again to the religious sisters located at the site of the old orphanage and ask for more specifics about the school's practices a hundred years earlier. They asked for details about how Jim Collins had been admitted and discharged from St. Agnes Home and School. They asked whether the Catholic institution might have accepted children of other faiths—one of whom, a Jewish child, could perhaps have been confused with another boy named Jim Collins and assumed his identity.

The reply, from the archives office of the Dominican Sisters of Sparkill, offered some important clues. The archivist elaborated on the school's policies and wrote that she was not aware of restrictions on who the orphanage took in. And she added that Jim and his siblings had been committed by the New York Society for the Prevention of Cruelty to Children, a step usually taken "if parents were deceased or unable to care for them."

This was news to Alice, Gerry, and their brothers, who had not known how the Collins children were committed to the orphanage. It opened up a new window into their father's early life.

Founded in 1875, the New York Society for the Prevention of Cruelty to Children was the first child protective agency in the world. (Notably, it was founded nearly ten years *after* the American Society for the Prevention of Cruelty to Animals.) Alice wrote to the organization and got back a lengthy summary of the files in its archives, which helped her piece together a rough idea of her father's first year of life. Jim and his siblings were very young when the society was first contacted in 1914 by a woman who said the children were "in need of everything." Society officers went to the top-floor apartment of an address on Southern Boulevard in the Bronx, and children answered to their knock. Their mother was sick, the children told the strangers, and had told them not to let anyone in. The officers left.

Not long after that, when baby Jim was nine months old, the children's mother died, and the rest of the file, consisting of the accounts of various "informants" and an official from the city's Department of Public Charities, is a confusing timeline of turbulence for the three Collins children. They were so little. The eldest, Kitty, was just four. At various times they were being cared for by three different women, at least two of whom were aunts. Various claims were thrown around and sometimes contradicted or walked back: that their father, John Josef, was "intemperate," that he wasn't, that the parents had separated. Money was tight, space was tight; the children were moved around and separated from one another. One relative, whose husband worked as a city street paver, explained that she was caring for the two older Collins siblings in addition to three of her own, without any financial assistance from their father. This could not continue. John Josef, then working as a "rigger's helper," earning twelve dollars a week, met with a society officer and said he intended to have his children "placed." At a court hearing, John Josef agreed to pay one dollar a week per child and have the children go into custody of the convent. They were separated, with the two boys going to the orphanage in Rockland County and their sister placed in a Catholic institution in Manhattan.

As the years went on, the society kept track of the children and of their father, who would never again be a family. Within a few years of relinquishing custody, John Josef was said to be in bad health, and an officer visiting his "poorly furnished but clean" apartment found him "in very delicate physical condition." He'd begun working as a longshoreman, John Josef said, but his health made it difficult; "this had resulted in his being in arrears in his support payments and therefore unable to visit" his kids. At one point, he arranged for his daughter to be released to the care of an aunt who appeared relatively better off, but the boys had to remain at the orphanage. At another point, John Josef visited one of his sisters to ask to borrow a few dollars.

Alice took note of several things in the file. First, she noted that her father's parents had apparently separated, which seemed unusual for Catholic immigrants at the time, particularly right around the birth of a baby. Why would they separate, and why at that moment? Looking at the dates, she was struck again by the fact that her father was given away exactly a week before he turned one. "His father didn't even keep him around until he had a birthday," she told me, touching her chest. The last thing she noticed was that, as far as she could tell, Jim and his siblings appear to have been transferred quickly from the care of their father to the Dominican sisters, lessening the possibility that Jim's identity had been mixed up by the organization that had taken him from his troubled home.

But that still left the theory of a mix-up at the orphanage. It seemed unlikely that Jim's older brother, John, wouldn't have noticed or said anything if his baby brother's identity were confused, but it was a possibility, and Alice was committed to exploring every single one. Here again, the Collins kids were butting up against the passage of time and the limitations of the paper records. They did not know a whole lot about their dad's time at St. Agnes, in part because Jim didn't talk about it much. "If only we had a time machine . . ." Gerry mused.

But one of the few stories Alice and Gerry had heard about their father's childhood was that once a week the orphans were brought to a classroom to meet with visitors. "Every Sunday, Dad had to dress up in his Sunday best and he would sit in a classroom with his hands folded on his desk and wait for a visitor," Gerry remembered her mother recounting. For nearly a decade, he never had one, at least not that he recalled. Then, one day when Jim was ten, he was told someone was there to see him. He was "terrified," Gerry said. "So afraid that he peed in his pants," Alice added. The man was very tall, "a six-footer," Jim would tell his children decades later, recounting the story with matter-of-fact delivery. "Do you know who that is?" Jim's older brother, John, asked. Jim did not.

"That's your father," John told him.

The tall man asked if there was anything Jim wanted. Jim asked for ice skates. Some time later, the tall man came back, carrying ice skates. And after that, John Josef did not come back, because he died, at Bellevue Hospital, of pneumonia and septicemia, in his early fifties. Jim had just turned twelve.

Alice had a no-stone-unturned mentality. How, she wondered, could she determine whether the baby placed at St. Agnes grew up into the man she and her siblings knew as their father? She hit on the idea that there might be a specialist who understood how faces change over time. She found a forensic artist and emailed the woman, recounting the story of Jim. "We have one photograph, purportedly taken of him, his father, and siblings immediately before the children were placed in orphanages," she wrote. "Is it even humanly possible to 'age' a photograph of an eleven-month-old baby enough to identify him as an adult?"

It was possible, the artist wrote back. It would cost fifty dollars.

Alice emailed her dad's only baby photograph, along with two pictures of him as an adult. The forensic artist wrote that she believed the baby and the man were one and the same. The two pictures showed the same domed forehead and earlobes, she wrote, the same mouth shape and expression, the same facial proportions. Looking at the portrait of baby Jim with his dad and siblings, she observed that Jim's big brother, John, "looks far more like his father than the baby does."

The orphanage theory had always seemed thin, and now Alice, the daughter of a skilled card player, concluded that few gamblers would give it good odds—especially not while there was another moment in time that seemed like a better bet.

* * *

Who was Jim Collins? How did he come to be?

The power of the DNA revelation is that it provokes questions about the past, which seems suddenly not like the past at all. It makes you question fundamental truths. It provokes an accounting—of a life in an orphanage, or a time when a couple was having trouble conceiving, or when a young unmarried woman went off for a mysterious vacation and returned many months later. It invites revisions to things we long ago analyzed and incorporated into our personal narratives. It suggests the past is never over, but a living thing that can be amended. How do you make sense of a life that starts with a mystery? How do you tell your story without a beginning?

There's a strange circularity to Jim's time at the orphanage. Admitted a week before his first birthday, he was discharged a week before his fifteenth. He had not finished high school. No one knows why he left when he did, but according to records, he went to live with his brother, John, who was only a little older than him. And this period of time, with two teenage boys attempting to live on their own, began the chapter Jim liked to call his "misspent youth."

He collected money for an illegal gambling ring. He made money playing pool and poker and craps. "He developed a math system—I couldn't explain how it worked—for taking bets for horse races," his son Bill Collins told me. "He told me he did so well with it that eventually the bookies in New York would stop taking his bets."

Adulthood and the army seem to have mellowed Jim. His kids knew he'd had a difficult childhood, but somehow, Gerry told me, it had not made him bitter. "He wasn't around a lot but he was soft and gentle and very easygoing," said Brian Collins, who could not explain how a man with such tough beginnings emerged with his sweetness intact. Jim valued family; without much of his own, he'd adopted his wife's family, calling Alice the elder's mother "mom" and coming to adore his wife's sisters. "My mother was close to her family," Alice told me, "but with Dad it was something almost transcendental."

And Jim worked hard. With seven kids, money was often tight. In his job as a correctional officer in a state prison in Soledad, California, after twenty years in the army, Jim worked double shifts and holidays. The few times he got angry were in arguments with his wife over money, one Collins brother told me. "It was always living paycheck to paycheck," another said. The Collins family lived in a three-bedroom, two-bath house in Salinas, California, all nine of them crammed into 1,250 square feet. The two eldest boys slept on a couch with a trundle bed that barely fit into a tiny connecting room intended as a den, and when Alice (no. 3 in the lineup) and Bill (no. 4) got into fights, Alice would run laps through the house, busting through the accordion door to that room in her attempts to get away. In the living room, their parents "had their chairs and they were sacrosanct and you did not sit in them," Gerry said one day when Alice and I were videoconferencing with her from Alice's sewing room. "You did not!" Alice echoed. "You got a seat and it was your seat unless you got up."

The kids had their chores and every Saturday they did a full house-cleaning, top to bottom. They made their own lunches, and made their mother's breakfast every morning and served it to her in bed, and Alice told me she made dinner every night from the time she was ten. If the Collins kids were united in their affection for their father, they were also unified in describing their mother as difficult and overbearing. Jim Collins and Alice Nisbet were both older for their era when they married. He was thirty, and she was twenty-seven. What drew these two very different people together none of the Collins kids could quite say; their togetherness was simply a given. "He loved watching her," his grand-daughter Cherie Collins says. "What it was that they had I don't know, but there was a spark." And later, when Alice the elder was sick for the last fifteen years of her life, partially blind and mostly bedridden after breaking her leg at the age of sixty, Jim was his wife's full-time nurse.

Jim Collins had qualities that often endeared him to people. At the prison, he was clear-eyed about the inmates, but he treated them with respect, and that seems to have gone far. Bill recalls visiting his dad's workplace for a college project, and his dad introducing him to some of the prisoners. "They were saying, 'Don't worry about your dad, we'll take care of him,'" Bill recalls.

Jim was a little socially awkward. He could ask weird questions, and mangle his words, and be slow and deliberative in his thinking. "My father's favorite thing to do when he came home from work was to sit in his chair and read the encyclopedia," Alice says. (It was this habit that led to one of her parents' bigger fights, when Jim discovered a hefty phone bill that his wife had hidden in one of the volumes.) He did the neighbors' taxes for free because he just loved doing taxes. Cherie talks fondly of trips she took with her grandfather to the military commissary, where Jim would pull out coupons and work his double points to see how cheaply he could buy cartons of Jell-O. Cherie told me her grandfather used to go dumpster-diving behind his apartment to find old toys, which he'd clean up and give to his grandchildren. I stopped her when she said this to make sure I was hearing right, since old toys from dumpsters struck me as kind of gross, but Cherie insisted that "it was very sweet." Her little sister got a favorite bunny that way. The thing is, Jim never got toys as a kid. One year, he got an orange for Christmas. "The way he was raised, you didn't throw away a toy," Cherie said.

When Jim got older, he began to talk more about his life. One day in 1993, when he was visiting his daughter Gerry and her husband, Roger, they pulled out a video camera and taped him at the kitchen table. When I visited Alice, she showed me this grainy video, which featured her dad telling stories about the time he and a colleague at the prison were ambushed by a group of inmates hiding in a kitchen, and the time in the army in Germany when he pretended he didn't know how to play pinochle and managed to win a thousand dollars in an epic night of gambling at the officers' club. Jim explained that he had learned the rudiments of pinochle as a kid, and then, at great length and with mounting excitement, he explained a series of complex maneuvers in exquisite detail, playing out a lengthy game of double-deck pinochle in his head, with Gerry as his partner. "All you have to control is two suits," he said. "That sounds strange but I'll tell you how you do it." He talked of tricks and melding and round houses. He explained how he could infer the cards in his partner's hand by the ones she had already played. He switched perspective to describe how an opponent could counter his moves. He went on and on, grinning at the salient points, connecting the strands of this imaginary game as if he were following an elaborate chart only he could see. "You add your six queens, you add your eight kinds, you add each marriage, a king and queen, and you add the marriage, and trump is four . . ."

Jim Collins at age eighty had receding salt-and-pepper hair neatly combed back from his forehead and a pair of thick-framed glasses. He had the same round face, mischievous smile, and dimples he'd had in that photograph taken when he was a baby on his father's lap. He had an old New York accent: "oil" was "erl," "cards" was "cahds." Sometimes he put his finger to his nose, as if signaling his interlocuter to pay attention. Other times, he smiled and rubbed his hands together, with the air of imparting a secret, or spread his hands out wide to demonstrate the hopelessness of being outwitted by a cunning opponent. Morning sunlight streamed in through the window, making stripes on his forearm, and I let his words wash over me as if they were in another language. "Follow me?" he'd say, after explaining something, as if everyone else could see this complex imaginary game, too. "You got that?" "See?"

"No, I don't see that," said Alice, who was sitting next to me, watching her father on the screen. I turned to look at her and she shrugged, and I was relieved I wasn't the only one.

12

THE SIMPLEST EXPLANATION

Alice and Gerry talked and emailed constantly as 2012 rolled into 2013 and they chased down lead after lead. They were beginning to think only one explanation made sense.

If the mystery of Jim wasn't buried in the old country, and if it wasn't confusion around the identity of a baby who'd been relinquished to a child protection agency and then to an orphanage, could something have happened right around the time Jim was born, on September 23, 1913?

Early on, before Alice stumbled across the clue that changed the way they thought about everything, Gerry had come up with a scenario that could explain how a Jewish baby from eastern Europe could wind up with Irish Catholic parents. Perhaps their grandmother Katie Collins, giving birth in her apartment with the help of a midwife, had birthed a stillborn child. Elsewhere in the Bronx, Gerry imagined, a Jewish woman had had a baby but then she died—and somehow, the two families had found out about each other. Could a bereaved John Josef and Katie Collins have taken in that other woman's baby as their own?

But then, Alice found the clue, and everything changed. The funny thing was, she'd had it in her possession since she'd first started her genealogical research on her dad. Jim's birth certificate had arrived in the mail back in 1998, shortly before Jim had died, and Alice had looked at his birth day and year, and looked at his parents' names, but she'd failed to notice one key detail. This part of the story reminds me of that psychology experiment in which subjects asked to focus on a video of a basketball game fail to notice the person walking past in a gorilla suit—sometimes you don't see something because you're not looking for it. Alice finally saw the detail fourteen years later, when she was in the sewing room/office of her house in Vancouver, mired in the mystery of who her father was, and she studied his birth certificate again. The detail was right there, stamped on the bottom, and again, in looping script, on the line next to "place of birth."

It said: Fordham Hospital. In other words, Jim had not been born at home, as everyone in the family had assumed.

Fordham Hospital during the era when Jim Collins was born

She told Gerry, and they discussed the implications of this information. Gerry thought one explanation that might fit the facts was an update of her baby-dying-in-childbirth theory. Maybe Katie Collins's baby died in the hospital, and the woman in the next bed died, and voila, a bargain was struck. She knew it sounded "cockamamie," but "fact can be stranger than fiction," and it was true, the facts were already proving so strange.

But more likely, Gerry and Alice both thought as they digested the news, there was another, less complicated, explanation. "That's what we did at work," Alice says of her old job, when she needed the most direct approaches to wrangle phone trees or online enrollments. "You work on the simplest way of doing something." Of course, the theory the sisters came up with seemed simple and obvious only because the more prosaic explanations that people like CeCe Moore see much more often—an affair, an adoption—had already been eliminated. Even Moore would later say that the very first time she heard this theory, she thought it sounded too much like a soap opera to be true.

* * *

Alice did not know much about Fordham Hospital in the Bronx. It had closed in 1976 and eventually became a parking lot, and neither she nor Gerry—doing Internet research and calling around to places like the Bronx County Historical Society—could find any additional records pertaining to their dad's birth. What Alice would come to know later, when she began to research childbirth in America, was that 1913 was truly early days for maternity wards. As the childbirth historian Judith Walzer Leavitt told me, it was not until 1938 that even half of American births were taking place in hospitals.

Before the turn of the twentieth century, Leavitt says, "the ideal way to give birth was at home with family around, and a doctor coming in and helping, maybe a doctor and a midwife." Only a handful of women gave birth in hospitals—the poorest and most desperate urban women, sometimes unmarried and facing wholesale rejection from their families, or women with special medical conditions but without enough money to afford private doctors. But because notions around childbirth were starting to change in the early decades of the twentieth century, it's hard to say why Katie Collins chose to give birth at Fordham, a public hospital that had opened in 1892 and was part of a group of allied public hospitals run by a reform-minded board of trustees. Fordham and other public hospitals offered free care to the poor, and in 1913 they would have been serving lower-working-class and immigrant populations. Jim was born as the idea of hospital deliveries was becoming less unusual. As obstetricians advocated for their medical specialty, they promoted moving deliveries into the more controlled hospital setting, Leavitt says, even though hospital births weren't necessarily safer yet.

Before John Josef and Katie Collins started their family, Fordham had been located in a cramped and out-of-the-way location, but by 1907, it had a new facility "in the heart of the swarming population of the Bronx," one chaplain wrote. The new hospital, erected at a cost of about $600,000, was "thoroughly modern in its plan and equipment," a publication gushed, "ideally situated on Southern Boulevard, facing the beautiful Botanical and Zoological Parks." The huge wards were painted white and filled with natural light, and patients sometimes convalesced in rocking chairs. Doctors wore suits and ties, while nurses wore crisp white uniforms with voluminous shoulders and full skirts. Fordham's new location was actually a straight shot down Southern Boulevard from where John Josef and Katie were living in 1913, in a new five-story building that had running water and may even have had steam heat. They were living right by several busy intersections and, for five cents, could hop on the subway at the nearby Interborough Rapid Transit station. The Collinses would have been surrounded by first- and second-generation immigrants—fellow Irish but also Germans, Italians, and people referred to in the census as "Russian Yiddish." They had married only a few years earlier, in 1909, when they'd both been in the country less than a decade, and quickly started a family. By the time Katie went into labor with her third child, she might have walked to Fordham, as some laboring women did in this era, or perhaps she took the hospital's newfangled "automobile ambulance."

The year little Jim was born, the population of the Bronx was six hundred thousand, and to keep pace with growing demand for the borough's only municipal hospital, Fordham had just built another ward to separate medical from maternity cases, rented a nearby house to accommodate its growing staff, and would clearly need another expansion very soon, its trustees declared. Its nurses would likely have been specially trained in how to care for the inhabitants of its maternity ward, which was small but growing by the year. According to pamphlets from the Bellevue Training School for Nurses, which appears to have been serving Fordham and several other hospitals in this era, nurses learned about the stages of labor, Cesarean sections, and post-labor complications.

They learned about newborns—caring for their eyes and for the dressings on their umbilical cords, the ins and outs of breast-feeding and infant formula. They seemed to offer poor women state-of-the-art care.

And yet, when I spoke with historian Sandra Opdycke, who wrote a book on the city's public hospitals called *No One Was Turned Away*, she told me Fordham was very likely being largely staffed by students still learning their craft. "Pupil nurses" provided the bulk of the nursing force and many of the interactions with patients. Old pamphlets from the Bellevue Training School indicate these pupils worked twelve-hour shifts every single day except for "an afternoon during the week, and half of Sunday." They were paid in food, lodging, and education, with the tiniest of stipends to cover uniforms and textbooks. In between all this, they had lectures and exams—though Opdycke told me she never could figure out how these students fit their studies around their work shifts. Under the nursing staff, which was typically spread thin, there would have been "hospital helpers," poorly paid, poorly trained, and poorly regarded staff with fantastically high turnover, who would also have helped with the newborn babies, Opdycke says. In other words, the people in charge of the hospital experience of someone like Katie Collins and her baby were likely tired, overworked, inexperienced, and underpaid.

Nearby Fordham University was not formally connected to the hospital, but the two institutions had a relationship, and university medical students trained there. In 1911, a promising young man received his medical degree from Fordham University. His name was Jerome McSweeney, and he graduated with the Dean's Prize, for the highest average throughout the entire course. Later, as an officer of the medical reserve corps during World War I, he'd be sent to Army Medical School in Washington, DC, return and buy a house on Bathgate Avenue in the Bronx, eventually open an office on Park Avenue, and die at the age of fifty-eight, still living in the borough. But in September 1913, McSweeney was a young doctor working in the Department of Gynecology at Fordham Hospital, which was where his path, along with those of a number of overworked nurse-trainees in peaked white caps, would intersect with that of an Irish immigrant giving birth to her third child.

* * *

By 2013, Alice and Gerry were buried in the genetic segments they hoped to follow like breadcrumbs on the trail of history. The sheer volume of their data was astonishing.

Over the years Alice and Gerry were working their case, the databases yielded a total of 6,912 DNA cousins for the seven Collins siblings, who altogether had 311,467 "segment matches"—that is, segments along the chromosomes that overlapped with those of the Collins siblings. Each one of these slivers of genetic material was a potential clue, as were the surnames associated with the matches, all 14,182 of them. The sisters imported all this data into vast Excel spreadsheets that showed an array of information along rows and columns of cells, including the number and length of the overlapping chromosome segments, and the locations of those segments along the chromosomes.

For instance, for a particular genetic segment match along chromosome 16, one cell might indicate the starting point was "11376029" and the end point was "17065473"—a detail of profound granularity that would mean nothing to you or me, but that Alice understood could hold the key to understanding who her father was, if she could just deduce what these numbers were trying to tell her. Sorting the data various ways, Alice and Gerry went down the list, Gerry following Alice's methodical approach and sometimes suggesting refinements to make their search more efficient. They examined match after match, chromosome after chromosome. Separated by 1,700 miles, the two sisters sent each other updates by email when they found promising relatives to pursue, and every day, Alice searched the spit databases to see if new cousins had turned up, and when they had, she downloaded their data and manually copied-and-pasted it to her spreadsheets. She tracked big trends and granularities, noticing what countries and surnames were showing up over and over, trying to make sense of it all. "If you're look-ing for relatives, you're looking for patterns," she told me.

Keeping all this data on spreadsheets became so unwieldy that Alice and Gerry's brother Jim the younger, the retired software engi-neer, created an iPad app to manage it all. Within the app, Alice could manipulate the data in different ways, searching and sorting by length and by chromosome, and in a notes field beside each DNA cousin, she kept track of pertinent information, like the names of the villages a cousin's grandparents hailed from or the fact that one cousin matched another along the eighth chromosome. She asked her brother to add cer-tain functions to the app, and in those instances when automation failed her, she used brute force—spending weeks, for example, just moving data from one field to another so that she could search it.

Meanwhile, to gather all this data, the sisters were reaching out to DNA cousins to ask them to share their genomes and other informa-tion. Over time, they sent at least a thousand such requests through 23andMe. While some people ignored these messages, plenty were intrigued by the mystery the sisters told them about, and eager to offer Alice and Gerry genetic genealogy techniques, or to scour the Internet for clues into the family trees of other relatives. Some became friends. The long history of marriage among Ashkenazi Jews meant that these helpful cousins were almost always not as closely related as the data-bases suggested they were, which made Alice and Gerry's job much more difficult. "There are only a few people I know of who really know how to do a Jewish unknown parent search—it's just incredibly mud-died by endogamy," says Jennifer Mendelsohn, who runs a Facebook group for Jewish genetic genealogy. "If you pull any two Jews off the street, the chances are pretty high that they're going to share some seg-ments." For example, Alice shows up as related to Bennett Greenspan in FamilyTreeDNA's database; I show up as related to his business partner, Max Blankfeld. This meant that not only were many of the sisters' sup-posed relatives essentially false ones, reflecting distant ancestry many times over, but they also had many more people to comb through. And that's not all of the challenges Alice and Gerry faced. The search for ancestral records is difficult in places where many Jews historically lived. "There's this incredible complication of provenance because borders changed so much," Mendelsohn says. "The place where my grandfather was born was part of the Austro-Hungarian Empire, and then it became Poland, and then Ukraine."

In the beginning, when she didn't know any better, "I'd look at people who were probably related to us through Moses," Alice says. And yet, once upon a time, far enough back, the Collinses and these distant cousins *had* descended from the same people, had shared soup and heartache, had slept in the same bed. A 2014 study in the journal *Nature Communications* found that all modern-day Ashkenazi Jews are likely descended from what researchers describe as a population bottleneck in the late medieval times. The population was very small, perhaps as few as 350 people, but more likely somewhere in the thousands. Something like ten million Ashkenazi Jews descend from this tiny group, which lived, according to several studies, around twenty to thirty generations ago. After centuries, Alice and Gerry were connecting again with their people through the technology of DNA testing, and although the genetic material of these distant cousins might not help the Collins family solve their case, perhaps their research efforts might.

* * *

Alice's problem was complicated enough that she knew using autosomal DNA alone might not be enough. So, even as she and Gerry worked through those thousands of database cousins, she convinced her brother Bill to take a Y-DNA test through FamilyTreeDNA, which offers a database of matching customers. When the results came back, she began something CeCe Moore calls "fishing in the Y chromosome."

The idea was to find other men closely related to Bill Collins along the male sex chromosome and look at their surnames. This works because a father typically gives two things to his son that are of interest to genealogists: a last name and a Y chromosome, which stays virtually unchanged as it gets passed down. Within many populations, the Y chromosome and the surname can be associated going back many generations, with occasional interruptions for things like adoptions and those "non-paternity events." So, a man fishing for the identity of his father can look at his matches in a Y-DNA database, hook a likely surname—let's say fourteen instances of the last name "Ogle" show up among men with similar Y chromosomes—and then go back and search among close autosomal DNA matches for folks with "Ogle" in their family tree. If the method works, he can trace his way into the right branch of the Ogle family.

Alice figured this technique was worth a shot, even though she knew that once again, the unique history of the Ashkenazi Jewish people might pose a challenge. Most Jews in eastern Europe only acquired fixed surnames starting at the end of the eighteenth century, in response to laws forcing them to adopt them. As a result, says Moore, Ashkenazi Jews "with the same direct paternal line two hundred to three hundred years ago are unlikely to carry the same surname now." Yet, incredibly, when Alice examined Bill's results, she found a man with a very close match to his Y chromosome. The man was unreachable, but Moore somehow figured out his full name, and Alice's brother Jim managed to find the man's aunt listed online at a synagogue. Alice got in touch with the synagogue's rabbi, who connected her to a relative, who connected her to the aunt, who agreed to take an autosomal DNA test to see if she was related to the Collins siblings. Alice bought a kit and had it sent, and the woman spit in the tube, and they waited and—

There was no match. Well, they were related, of course, in the way that Ashkenazi Jews are often related—"pull any two Jews off the street," as Jennifer Mendelsohn would say—but very distantly, many generations back.

It was disappointing, but Alice plodded on. One helpful DNA cousin suggested she try another helpful DNA cousin, who passed her on to a third helpful DNA cousin, who happened to be a genealogy scholar, who roped in another scholar, and the two men began brainstorming ideas for how to help Alice and Gerry. Finally, one of them suggested to Alice that the answer to her mystery might lie in the last name "Cohen." There were a few reasons why this idea made sense, including the notion that if the nurses at Fordham Hospital had relied on an alphabetical system to organize their babies, they might have confused a little Collins with a little Cohen. Alice reached out to genetic genealogist Gaye Sherman Tannenbaum, who was able to look at a copy of the New York City Birth Index of 1913 and find one Cohen who fit: a Seymour Cohen, also born in the Bronx, and on the same day as Jim Collins, September 23, 1913.

Alice began to do research on the family of that little boy. She found a birth announcement and figured out where Seymour and his parents had lived, not too far in the Bronx from John Josef and Katie and their children. She traced where they'd come from in Europe. Was this Cohen family her family, too, the answer to so much hard work and searching? She reached out to Moore, who looked at Bill's Y-DNA results and noticed a fairly good Cohen match, which added extra credence to the theory. "It just felt right," Alice said.

It felt so right that Alice and Gerry decided to announce their news to the family. Alice prepared an email. It was May 2013, just under a year since she'd sent away for that first AncestryDNA test, and while that seemed like a long period to be immersed in this consuming and painstaking work, it was, she felt, actually pretty short, given the degree of difficulty of what she and Gerry were trying to do.

"Our search for Dad's parents is most likely over," she wrote, and revealed to her siblings their Cohen grandparents' names. It was an astonishingly gratifying moment.

To confirm what they already knew was true, Alice and Gerry needed to compare their DNA against someone from that little boy's family, so helpful DNA cousins fanned off across the Internet and helped track down a professor in North Carolina who was a descendant of Seymour Cohen's sister. Alice wrote to her and asked if the woman would be willing to help them solve an incredible genetic detective case, and the woman said she'd be delighted. Alice paid for the kit and had it sent, and the woman contributed something completely free and absolutely priceless. She sent it in, they waited, and—

Well, they were related a little bit, if you went back far enough. "Were we ever wrong," Alice says.

13

THE AMERICAN FAMILY

Stephanie Talley learned she was donor-conceived when she was twelve, but it didn't mean too much to her then. It wasn't till later, when she got to the stage when "your parents are suddenly embarrassing and not cool—that's when I started to wonder about that person who contributed 50 percent of my body," she says. "I used to look in the mirror and try to subtract my mom." Her parents divorced when she was a teenager, and after that she wasn't close with her dad, who mostly disappeared from her life. Over time, Stephanie romanticized her anonymous donor, imagining him as an "artistic beatnik guy" responsible for her love of the arts and photography, and wanting to find him. But she couldn't get very far until consumer genomics came along.

Stephanie started testing in 2012, and got her first big break two years later when a half-brother showed up in the relative database. Then she found another half-brother and a half-sister, all of them donor children, and they began to get to know each other. She was using genetic genealogy methods to piece together her donor's identity, and was about to solve her case in 2016 when her biological father showed up in the AncestryDNA database. By then, there were five donor-conceived siblings, and together they crafted a letter to send him over Facebook Messenger. They went back and forth, fine-tuning the tone. Talley told me she understands why so many seekers fear rejection when they reach out: "I was so worried about being good enough to be wanted."

But the donor did not shut them down. He had raised two daughters, and he felt the decision about whether to know these donor children—his daughters' half-siblings—was more theirs than his. He talked to them, and they were open to it, so he and they and some other members of his family agreed to meet up in New York with his donor-conceived children. Stephanie came up from the Tampa Bay area and was so anxious before the meeting that she threw up. What struck her during this visit were the commonalities that genetics seemed to have bestowed on her and her half-siblings, beyond even their physical likenesses: shared sarcasm, outgoing temperaments, artistic inclinations. It turned out her teenage imaginings had been right: Her donor father has a degree in fine art and, after retiring from his career as a civil servant, now works in ceramics. "I feel whole," she confessed over dinner with her biological family, tears springing to her eyes as she looked at one of her donor father's daughters. Her half-sister teared up, too, and they hugged.

But the story didn't end there, because the siblings kept showing up in the DNA databases. When I first interviewed Stephanie during the summer of 2018, she told me her donor father had a total of fifteen biological children "so far," including the two he'd raised. When we spoke again less than a year later, there were nineteen, the last having shown up within the previous forty-eight hours. I talked to their donor father, Robert Jenkinson, and he told me he figured he'd about neared the full tally of his donor-conceived children, although "I wouldn't be surprised if there's another one out there." (Months later, the count had gone up to twenty-three.) Most, though not all, of the donor-conceived siblings were forging relationships within this new family. A few had spent Thanksgivings with Jenkinson and his daughters, and gone camping with them at Virginia's Shenandoah National Park. One of Stephanie's half-sisters had come down to visit her in Florida. For an only child, Stephanie told me, having this abundance of family in her thirties was wonderful, and seeing some of her own qualities reflected back in her siblings had given her "permission to be myself more." And yet, aware of how many other donor-conceived people have had bad experiences in their search for genetic family, Stephanie says, "it's hard for me to completely bathe in how good it is." She told me she feels as if she's won some sort of seeker's lottery and worries about trumpeting her good fortune, not wanting to belittle other people's experiences or gloss over the complications of being donor-conceived.

Some donor-conceived people are deeply troubled by aspects of sperm donation, including instances in which people are not told the truth of their origins until later in life—or until DNA testing uncovers it. As with parents of adoptees, parents of donor-conceived children are generally more open with their kids about how they came into the world than they were decades ago, when, as a piece in the journal *The New Atlantis* explains, the shame of male infertility and the "stigma of illegitimacy" encouraged parents to keep the practice quiet, including from their own children. (The tradition of secrecy goes back to the 1880s, when, in one of the first documented uses of artificial insemination by donor, a doctor sedated and impregnated his patient with the sperm of a good-looking medical student, without telling her or her husband.) Once upon a time it was not uncommon for doctors to report mixing donor sperm with the sperm of the intended father in order to "circumvent legal or psychological difficulties" by keeping the child's paternity ambiguous. This was sometimes known as "confused artificial insemination," and I found a 1990 paper stating, less than reassuringly, that it had by then "almost died out." Plenty of people in their forties, fifties, and sixties are just discovering now how they came into being, and often their parents are no longer alive to have conversations about what they knew and how they felt about it.

But even when donor-conceived people are told the truth of their paternity in childhood, some of them describe the way it is practiced in the United States as "unethical" and have found the truth of their conception destabilizing. "Conversations about this topic happen about us—without us," Joanna Rose, an activist for the rights of the donor-conceived, has said. Stephanie Talley told me she's conflicted by the technology that brought her into being; on the one hand, practices like sperm and egg donation enable gay couples, single women, and those with fertility problems to have children, and she supports that. On the other hand, these practices seem to place a parent's need for a biological connection to their child by means of assisted reproduction above the child's need for that same connection to the person who contributed half of their DNA.

Donor-conceived people have been seeking their genetic families for a long time. In 2000, Wendy Kramer, urged on by her donor-conceived son, started a Yahoo group that became the Donor Sibling Registry, a nonprofit established to help people conceived by sperm, egg, and embryo donation make contact with their genetic families. Kramer's organization also conducts research and advocates for the rights of the donor-conceived, as well as for better regulation of the industry. The registry has more than sixty-five thousand people in paying family memberships in more than one hundred countries, attesting to the deep desire on the part of the donor-conceived—as well, as in some cases, the donors themselves—to seek out their biological kin. But this precise moment, in which so many people are discovering who their genetic fathers are through DNA testing, is at times exacerbating tensions between people who are strangers in all but genes. Most donors gave sperm with the understanding that they'd remain anonymous, and Stephanie has seen seekers receive legal documents, warning them that additional attempts to contact donors will be considered harassment. (In one case, reported in the *New York Times*, an Oregon mother who contacted her daughter's biological grandmother after finding her on 23andMe was threatened with $20,000 in penalties by the sperm bank she'd used to conceive her child.)

Genetic testing is also showcasing what many see as inadequate regulation of the industry that brought them into being. Some want limits on how many children a donor can produce. In contrast with a number of countries, there are no legal limits to this in the United States, though the American Society for Reproductive Medicine, a professional society, recommends no more than "25 pregnancies per sperm donor in a population of 800,000" to prevent what it calls "inadvertent consanguinity"—as in half-siblings having a child together. Within the Donor Sibling Registry there are a number of groups with at least 100 genetic half-siblings, with one group hovering around 200, Kramer told me. "Inadvertent consanguinity" is a real concern for the donor-conceived, especially in cases where many live in the same city; Kramer has said she advises people "to memorize their sperm or egg bank name and donor number, and to share that information with potential dates."

The DNA age is also bringing to light more cases in which fertility doctors appear to have secretly substituted their own sperm. There are cases like this from all over the United States and the world, but one of the more well-known involves Indiana doctor Donald Cline, who, according to the *The Atlantic*'s Sarah Zhang, told his patients in the 1980s that he was using medical residents as donors and limiting their contributions to a small number. Instead, *he* was the donor. At least sixty-one people say they've discovered Cline is their biological father from genetic tests, the *New York Times* has reported. He pled guilty to obstruction of justice charges for initially denying the allegations, but received only a suspended sentence. Cases like this have led to lawsuits and calls for new laws, and in 2019, Indiana passed a fertility-fraud law initiated by several of Cline's donor children.

If assisted reproductive technologies have changed how we can make nuclear families, it has taken DNA testing to demonstrate how far beyond "nuclear" the definition of "family" can extend. How do you adapt language to account for the kind of family that someone like Stephanie Talley now finds herself in? People have invented new nomenclature for these relationships, with some in these expanded networks referring to genetic half-siblings as "halfies" or "sperm siblings" or "diblings" (for donor siblings). Stephanie, who dislikes the "arm's-length quality" that a term like "dibling" implies, and simply uses "brother" and "sister," struggled with what to call Robert Jenkinson, feeling "donor" was sterile and "dad" too loaded. They talked it over and she eventually arrived at "Pops."

As for Pops himself, he told me he never expected any of this; he couldn't foresee the future any more than the people who ran the fertility clinics could. He says he donated for about two years in his twenties, when he was newly married with a new daughter, looking for a job, with money tight. It was the 1970s. His wife at the time had given him the idea for how to make a little extra cash, at twenty-five dollars a contribution, if memory served. He never thought he'd meet a child he'd helped conceive; instead, he thought of it as helping someone else expand their family.

And then, several decades later, his daughters gave him an AncestryDNA kit as a Christmas gift, as happens in so many families, and he never realized it would give him anything beyond that little pie chart. ("Surprise!" his current wife said when they recounted this story.) Jenkinson says the biggest gift from this revelation has been the relationships forged between all his biological children. "I hope that continues after I'm gone," he told me.

When we spoke, Stephanie and two of her donor siblings had just returned from the camping trip with Pops and his wife and the two daughters Pops had raised. "It was cool, it was a family thing," she said; even so, she'd been so anxious during the trip that she'd thrown up again. Sometimes, she felt like she didn't deserve this newfound, uniquely modern family. She knew some donor-conceived people who could only dream of a camping trip like that. And on the other hand, she knew many people who were not donor-conceived and who took for granted the warm, loving relationships they enjoyed with their genetic families.

"I feel so privileged, but should I?" Stephanie asked. "Why should I feel like I don't deserve it?"

* * *

Even though it was clear that a long-ago baby named Seymour Cohen was not the solution to Alice and Gerry's puzzle, the sisters believed they'd been right in one crucial sense. Some baby, born so long ago that he almost certainly was no longer alive, was the missing piece, if they could only figure out who that baby was. At Fordham Hospital on or about September 23, 1913, their infant father had been switched with another, and the two had gone home with the wrong families. It had to be. The idea that their father had been switched at birth was the only theory they had left.

THE AMERICAN FAMILY

Alice being Alice, she started hammering away at proving this theory. She figured that the name of this baby was surely somewhere in old genealogical records. So, she began working with the New York City Birth Index of 1913. The Bronx in 1913 saw 14,679 births, and for Alice this meant 159 pages of baby names. Unfortunately, the names of these babies were not organized by what date in 1913 they'd been born; nor did the records specify which births took place in the hospital—that would have helped narrow it down. Instead, the names were alphabetized, which was thoroughly unhelpful, because Alice didn't know the name of the child she was looking for. Undaunted, she put all those baby names into a white binder, and then she set about combing through every single page, squinting at the old type, which was sometimes blurred into illegibility.

What Alice was looking for were a few demographic factors: a child, likely male, with a last name that sounded either Jewish or ethnically neutral, born within a day on either side of her father's birth date, September 23, 1913. Although Katie Collins had spent thirteen days in the hospital, not unusual for the time, Alice figured only those three days mattered because only just-born children were relevant. No mother who'd spent days holding her infant could mistake some other child for hers.

She searched methodically as best she could, marking all the names that interested her, which were a lot. She tracked the birth certificate numbers associated with these baby names, looking for ones close to her father's birth certificate number, figuring a baby born in geographic and temporal proximity to her dad would likely have a number close to his. Her white binder gradually became littered with turquoise stickies with names and numbers written on them.

Then, she began to play with her data. Over the previous year, Alice had collected an awful lot of family surnames from DNA cousins: so many Levines and Cohens and Friedmans, not to mention Slutzkys and Franks and more. Now, she attempted to see if the last names of any of the babies-of-interest from the index showed up within the pool of

209

surnames she'd collected. If a close DNA cousin had one of these surnames in his or her lineage, they might be related to the woman who'd
given birth to her father—back when he was intended for a name other
than Jim Collins.

About thirty-five names seemed especially promising, and Alice
made herself a spreadsheet of them, then searched systematically in
alphabetical order. She started with Appel, then Bain, then Bamason (or
was it Bamson? Or Ban_son, with an illegible letter?), Bohrm, Bresky,
Brody, Cherkunes, Dazey, Edelman, Efferenn . . . and all the way up the
alphabet to Paltzold. She was getting hits, but not on any close cousins.
She'd spent enough time in old censuses and the like to know that
officials sometimes had atrocious handwriting and a flagrant disregard
for original spellings, so she did wildcard searches, in case names were
misrecorded. Through the summer of 2013, Alice searched this way,
scrawling handwritten notes to herself on a copy of the spreadsheet
she'd printed out, writing down when a baby surname she was interested in showed up associated with people predicted to be DNA cousins.

And she got nowhere. It appeared to be another dead end.

* * *

Back in the early 1990s there was a pretty crummy movie starring
Brendan Fraser (remember him?) as a frozen caveman who gets thawed
by some high school kids and gets to live among us moderns, and that's
what I think of sometimes with respect to consumer genomics—all
those secrets frozen and brought back to life in a vastly different era.
These days, we place a high value on transparency and truth-telling, but
the truth as revealed by our saliva is at odds with certain historical high
stakes. Infertility, donor conception, adoption, extramarital affairs, and
interracial relationships once could have threatened a family's integrity
or its standing within a community. Some of those stigmas seem so
dated that you'd think these secrets couldn't hurt anymore, but they
do. Because, in fact, out of those secrets came people—people who

can sometimes still sense the shame that surrounded their hidden identities.

Genetic testing makes clear that the past was a foreign country. Once upon a time, as Rosario Castronovo discovered when he was researching his grandparents, Vermont convicted and imprisoned people for the crime of adultery. Once upon a time, young women who became pregnant when they were unmarried were sent to places like the Veil Maternity Hospitals in Missouri and Pennsylvania. Promised protection of their "reputation and social standing," the women gave birth, sometimes under assumed names, and typically never saw their babies again. I heard about one of those babies, now in her eighties, who spent years searching for the identity of her mother before DNA cracked the case.

Such late-in-life revelations have an extra poignancy. A Florida TV station reported on how DNA testing had for the first time brought together four elderly siblings, the product of a man who'd divorced his wife, changed his name, and started a new family. "To learn at this time in my life that I have a sister—and my sister's 102?!" said one of the men, grinning and clutching his cane, while the sister beside him smiled behind her walker. I talked to two sisters, both in their eighties, for whom DNA testing had become the means to reconnect after forty-five years of estrangement. They never addressed the reasons for their split—"We purposefully don't talk about any bad things, only good things," one of them told me—but this late-in-life reconciliation was deeply meaningful.

And in another case, the *New York Times* reported on a woman who was unmarried when she gave birth in a hospital in 1949. She was told her baby had died. In fact, that baby was alive and was adopted out; somebody, as CeCe Moore would say, was playing God. The mother was eighty-eight years old when she met her sixty-nine-year-old daughter. What do you make of such a story? That the loss is heartbreaking? Yes. That the revelation is a lucky one? Yes, that, too. The gratitude and the sorrow are woven together. DNA testing is a race against time.

Part of what interested me in talking to seekers was the question of how they processed the pain of the past: the unfinished business, the unanswered questions, the unresolved anger and shame and resentment and grief. And how did they make sense of lives with huge people-shaped holes in them? So often, the secrets themselves loom as a kind of negative space in those timelines: the teenage girl who went away for months, or the father who disappeared, or the relatives whose names are listed together on a ship's manifest, yet for some reason scattered in the new country. With the truth written into visibly overlapping DNA segments in commercial databases, we are all now left to make something out of those negative spaces.

"There are secrets people don't want to talk about and then they pass, and there's no way they can talk about it," Krista Driver says. Driver, a therapist now in her late forties, was adopted. When she met her biological mother at twenty, her mother told Driver who her father was, and Driver went on to meet this man and to believe he was her genetic father for twenty-six years. And then, in 2018, she took a DNA test, and that blew everything apart. When she figured out who her real biological father was, she texted her mother his name and asked, "Who is this?" Her mother never replied. "She hasn't stopped crying" since you texted, Krista's half-sister wrote her. When I spoke with Driver in the late winter of 2019, she told me her biological mother had passed away just three weeks earlier. She had taken her tears and explanations to the grave.

Driver told me she worries about the way the press focuses on happy reunions when they cover genetic testing surprises, failing to account for the emotional complexity of such situations. Driver is CEO of the nonprofit Mariposa Women & Family Center in Orange County, California, and since her own DNA revelation, she has started counseling groups for those who've experienced NPEs (which she defines as "not parent expected"). "There's nothing really in the mental health community addressing this," says Driver. She's seen others express frustration: "I went to a grief specialist, but they didn't really understand." Driver doesn't directly treat people who've experienced NPEs because her personal experience makes her too close to the topic, but she trains other counselors who do, and she says there's often a panoply of emotions accompanying the discovery that one or both of your parents are not who you thought. One is a particular grief at "losing a life that you didn't even *know* you didn't have," she says; she was herself flattened by this emotion after a reunion organized by her genetic father's family at which she saw four cousins who grew up less than a mile from her, and with whom she could have had lifelong relationships if only she'd known they existed. There's also sometimes anger at the parents who didn't tell the truth, and the ensuing guilt over being angry at one's parents.

And there's anxiety at the loss of identity. "Human beings have an instinctive need to connect with each other; we're pack animals," Driver says. "For somebody to suddenly be disconnected from their pack, from what they were connected to, it's like being unmoored, it's being untethered." Her words made me think of Jacqui Ochoa's description of the alienation she felt when she discovered her newfound siblings were not her genetic siblings, and the language Alice used when she learned her beloved cousin Pete Nolan was not her genetic cousin— she'd felt "adrift."

At its core, Driver says, a DNA revelation like this can create profound trauma, made all the more difficult to process at first because the cycle—from surprise to meeting new genetic family—is often so short. She told me she didn't spend a lot of time on a Facebook group she'd joined for those experiencing misattributed parentage because "a lot of people are so broken and angry." She's seen cases in which people in sobriety have relapsed after discovering genetic surprises.

A few other specialists have emerged to help people through the particular landscape of DNA revelations. Farther north of Driver in California, therapist Jodi Klugman-Rabb has been offering a treatment for what she terms Parental Identity Discovery. And across the country outside Charlottesville, Virginia, Brianne Kirkpatrick, founder of Watershed DNA, believes she is the only genetic counselor who helps people not only with the implications of medical genetic testing but with family surprises as well. She runs several secret and intentionally small Facebook groups for people who very often "just want to know that what's happened to them is normal."

Kirkpatrick preaches proactivity for the keepers of genetic secrets, both because people on the outside of a secret have a right to know who they are and because, in the age of DNA testing, they will eventually find out anyway. It's not a matter of *if*, but *when*. For parents who haven't yet told their children that they are, for instance, donor-conceived, she offers resources and helps them plan for these conversations. And for those who've discovered a surprise by way of a DNA database and ask, "Can I keep this a secret?" Kirkpatrick usually explains to them why this is impossible, even if they were to delete their account at MyHeritage and say nothing to anyone. Do they have children who might someday test? Do they have siblings who might? Time is the enemy of secret-keepers, because with time comes the inevitable revelation, one way or the other.

Kirkpatrick also works with the children of incest, a revelation sometimes made through DNA testing by adoptees, and one that can be incredibly painful. "Once you know it, you can't un-know it," she says. She prefers to use less stigmatizing language, describing these people as having "high ROH," which means "high runs of homozygosity," or stretches of identical DNA inherited from both parents. There's a tool on GEDmatch called "Are Your Parents Related?" that shows what this looks like, with long segments of identical DNA on both of a pair of chromosomes indicated by blue bars. In an information sheet Kirkpatrick produced for those with high ROH, she offers resources and explains the health implications—in some cases, being the child of incest can pose medical issues—and emphasizes that these results are not the fault of the person making the discovery, a point I saw made elsewhere. "You cannot control your own birth!" blogged one adoptee who wrote about confirming that his biological grandfather was also his biological father. "The important thing is to not let it control you or your life."

Among the groups she runs, Kirkpatrick started a group for women married to "birth fathers," men who discovered by testing that they had children they didn't know about. These women call themselves "birth wives." According to the accounts of these birth wives, there is "a very unusual dynamic that happens when it's an adult daughter coming into the life of her birth father," Kirkpatrick says. "The way the wives have described it, in some of the situations it turns into almost a romance," with the husbands secretly texting their daughters, buying them flowers, and primping before seeing them. "In some situations this is how GSA can happen," Kirkpatrick says, referring to something called "genetic sexual attraction," which has not been the subject of much research. GSA is not necessarily always acted upon, Kirkpatrick says, so it can't be equated with incest, but it can be troubling and painful for the people involved. Kirkpatrick says it has cropped up as an issue within the adoptee world in the past and worries it is now a looming matter of concern for the growing universe of people affected by DNA surprises.

In coming years, as history collides with the present, we may grapple with other issues stemming from such surprises, ones we can't even imagine now. Krista Driver argues that America needs to have a broader cultural conversation about DNA revelations: about how common they are, about the wide range of emotions that accompany them—not only on behalf of all the people who've already experienced their genetic surprises, but on behalf of those whose surprises are to come.

"If we start having a conversation now, and start taking the shame out of it, that day of reckoning for people may have less of a sting," Driver says. "We just need the past to catch up to where we are now."

* * *

Is it always better to know the truth? For seekers, I've found, the answer is almost always yes. CeCe Moore told me she's found the same thing, and it surprised her: "I really do expect more people to say I wish I never tested." That's despite the fact that, as Krista Driver and many others have pointed out, reunions aren't always "happy" ones. Sometimes, seekers are disappointed once they get to know the people they were looking for. Sometimes tensions arise over whose narrative of the past will prevail. It may take time for this to play out; I've checked back in with seekers after many months or a year and found their perspectives have shifted. For instance, the seeker I call Nancy, who got to know her father after taking a DNA test, began over time to doubt his account that he'd tried desperately to have a relationship with her when she was a child. Nancy said her father repeatedly avoiding discussing what she called "the pink elephant in the room, which was why the hell were you not in my life for the last forty-five years?"

And yet, Krista Driver says, "I wouldn't want to hold people back from finding their truth because they heard someone say how horrible their experience was." Because that truth is worth it. "I don't want to live with secrets. Secrets make you sick."

Brianne Kirkpatrick says she watched this sentiment echoed in one of her Facebook groups, when a man posted that he'd figured out his wife's biological father was someone other than she thought. "Should I tell her?" he asked. The people in the group who'd experienced similar surprises said yes. "There's a really big push from the NPE side of things that people are owed the truth about their genetic origins—even if it means a painful process," Kirkpatrick says.

To be clear, it's hard to extrapolate across the human spectrum from a pool of people who are invested in uncovering the truth. How much of this is the result of self-selection? "Maybe those are the ones I don't see because they don't pursue it any further," Moore says. And indeed, there are surely many people who receive surprising results after testing on a whim and don't immediately decide to devote hours upon hours to solving that mystery. In some cases, they might decline to look further because the news is destabilizing, whereas in others, they might not have the time or interest. Many testers are like Jessica Benson, a stay-at-home mom in her thirties living in Raleigh, North Carolina, who tested in 2014, hoping to find out more about her late father's side of the family. When her ethnicity results didn't line up with what she expected to find along her paternal side, she spoke with her dad's sister, who was equally mystified, and then threw up her hands.

"Somebody did something with someone they shouldn't have," Jessica concluded, and that's where she left it, until the truth came looking for her later. There are surely many more people with discrepancies like this lurking in their results, discrepancies that did not turn them into seekers.

Other things matter, too, in a person's experience of a revelation. Like the way the news unfolds for you. A philosophy student, writing in *The Guardian* about how genetic testing had uncovered that his father had lied about his ethnicity, rupturing his lifelong identity as an American Indian, resented "the insistence with which my aunt had pushed my sisters to take these tests," and the fact "that I couldn't discover who my father was on my own time." As psychologist David Brodzinsky puts it, "those who feel in control generally feel better about what's going on, and if you're doing the searching, you're the one in control." Some of the way we process these disruptions comes down to a sense of agency. It is one thing to be invested in a search, another to have the truth thrust upon you.

And of course, the stories of Laurie Pratt's father, who deleted his AncestryDNA account after she found him, and Jacqui Ochoa's half-siblings, who question the science behind DNA testing because her existence implies their mother abandoned a child, have demonstrated that it is a vastly different thing to be the *seekee*, especially when the news suggests something shameful or troubling about you or the parent who raised you. Genetic genealogist Debbie Kennett says there's a great need for research into the cultural and personal impacts of DNA testing and the secrets it uncovers. "We've got all these stories coming out, but certainly in our community, we're seeing a very one-sided view. We're always seeing the perspective of the people who are searching, and we never seem to hear about the effects on the people who are contacted," she says.

But for seekers, revelations about the past and how they came to be can offer context for otherwise inexplicable childhood memories. In some cases, they can answer lingering lifelong questions. "My entire

existence was a cognitive dissonance," one seeker told me. The truth col-
ors countless interactions, tiny and weighty, stretching back through a
life and before it. Ricki Lewis, who learned through genetic testing that
she'd been conceived by a sperm donor during an era when some doctors mixed sperm, has been waking up with old memories she hasn't
thought of in years, as if her subconscious is rewriting her childhood
in light of new facts. She looks back and wonders whether her father
knew he wasn't her genetic father, and if so, might this help explain the
depression he struggled with? She wonders if the fact that she was not
related to her father explains the almost instinctive affinity she felt as a
child for her mother's family, but lacked for her father's.

The need for certainty and clarity helps explain why people can
become obsessed with their genetic mysteries, their investment building during the painstaking process of discovery. Adoptee Linda Minten
told me what drove her search was "the truth, just the truth, just the
truth," beating the phrase like a drum. "I need the truth because I
can deal with that head on. It's all this underlying gray—secrets,
hiding—that stuff messes you up." The secrets were no foundation, she
said; they were like quicksand. Cultural psychologist Steven J. Heine
observes that uncertainty can be more difficult to handle than even sure
knowledge of bad news.

Knowing the truth is about claiming a complex, clear-eyed self.
Seekers told me they needed to know their own beginnings in order to
write the rest of their stories. A woman I'll call Sara told me about her
journey: about discovering through a spit test that her father was not
genetically her father, about unraveling that her genetic father was her
mother's first cousin, about calling her mother, who cried and said that
her encounter with this cousin had only just started coming back to her
and that what she remembered was that she had been raped.

Sara told me her upbringing had been deeply troubled and unstable, marked by her mother's emotional unavailability and abuse in the
family home. In that single phone call, things shifted. "It was the most
honest and real that I felt she had ever been with me in my whole life,
and I felt more connected to her after that whole conversation than I

ever felt," Sara said. "It was like all of a sudden, all of these pieces made complete sense." The news changed how Sara related to her mother, for whom she had more empathy and understanding. She encouraged her mother to see a therapist so that she could process what had happened.

Sara told me she was under no illusions that her family was somehow fixed. The truth was no magic balm. Her revelation had upended her mother's life, and she did not know how that would play out for her. But for Sara, at least, this tumultuous journey had been a profoundly good thing. Knowing the truth had freed her somehow, even though the truth itself was ugly.

* * *

In the summer of 2013, the Collins family held a reunion in San Francisco and celebrated the upcoming hundred-year anniversary of Jim Collins's birth. The siblings spent days together, talking about their dad and the mystery still at the center of his life. Alice and Gerry updated their brothers on all the work they'd done and all the techniques they'd tried, fielding questions and suggestions. Gerry had made beautiful stained glass Christmas ornaments for everyone, each a Jewish star of David with an Irish shamrock in the middle. Alice's son, Gidian, hung his from a button on his shirt, like a necklace.

The culmination of the reunion was a picnic in Marin County, at Mount Tamalpais. Alice remembers two things in particular from that reunion. She remembers the siblings discussing a theme that occupied many of their conversations during this time: that they would not and could not be bitter about what had happened to their dad. Without his case of mistaken identity, however precisely it had come about, Jim the elder would not have met Alice the elder, and the seven Collins siblings would not be. The vagaries of history had led them here, so they had to make peace with it.

The other thing Alice remembers is lying. She told her siblings she was certain that if she just kept plugging away, she could eventually find their biological grandparents. In truth, she felt no such certitude.

After the reunion, Alice went north to Washington State and kept working her case. She hoped that by Jim's birthday on September 23, she'd be able to solve it, but that was not to be. Instead, on the day after his birthday, she and her best friend, Kathy Long, set off for Europe. The birth index may have been blurry, and surnames and segments might mislead, but Alice knew the geographic origins of her DNA. Her closest genetic cousins had told her where their grandparents and great-grandparents had lived before they came to America, and she'd written down the names of all those towns. They went to Poland, Latvia, Lithuania, Estonia, and Russia.

It was an emotional trip for Alice, who little more than a year before had not known she had any connection to this part of the world. The friends visited a Jewish ghetto in Krakow and a memorial to Jews who'd lost their lives in Warsaw. When Alice and her siblings had lived in Germany in the mid-1950s, the reconstruction of the country that had murdered so many Jewish people had been the backdrop to their lives. Now, at the Warsaw Jewish memorial, Alice thought, *These are my people*. She began to cry.

In Lithuania, she and Kathy visited the capital city, Vilnius, and toured a museum that had been made out of a former Soviet Union KGB headquarters. Inside were photographs of Lithuanian people; terrible things had happened to some of them. And among the faces, Alice glimpsed a familiar one. It was a little boy who looked so much like her brother Brian had looked when he was young that her heart stopped. Alice stood there, riveted, unable to get over the idea that she'd come so far to see a face from home.

Alice flew back to Washington and kept plugging away at her detective work. She continued with her segment triangulation, day after day. She put the data from each new promising DNA cousin into the app Jim had made for her, and sorted and studied the results. She chased down leads, and was disappointed, and then did it again. In time, she and Gerry would buy at least twenty-one DNA test kits for themselves, relatives, and strangers suspected of being relations. And she tried to get close to her grandparents by homing in on where they'd once been. She watched a seminar intended for people researching ancestors affected by the Holocaust, and came away with a new technique, the fruits of which she showed me when I visited.

On her Mac, Alice navigated to a spot within Google Maps she called "ancestor pins." We watched as an image of the earth spun around to show the landscape of eastern Europe come alive with yellow dots. They were all over Belarus, Latvia, and Lithuania, but they clustered, like raindrops, over the Lithuanian city Vilnius, where she'd seen her brother's face.

She could see where her DNA was coming from, right there on the screen, but she was still wandering without a map. She was Alice in the wilderness.

14

YOUR GENES ARE NOT YOURS ALONE

In 2012, right around the time that Alice Collins Plebuch was trying out Ancestry's beta version of an autosomal DNA test, Yaniv Erlich was a fellow at the Whitehead Institute for Biomedical Research in Cambridge, Massachusetts. He was studying genetic privacy, a concept that fascinated him. But not everyone in his academic community was thrilled with his choice.

"Several people came to me and told me, 'Why even study this type of thing?'" Erlich told me over the telephone from his home in Ra'anana, Israel, in early 2019. In many ways, 2012 was a long time ago. "This is pre-Snowden, pre–Cambridge Analytica," says Erlich. It was before Russian bots on Facebook, before the election of Donald Trump. "2012 was just the dawn of big data," Erlich recalls. "People thought, *That's going to be great for everything.*"

Erlich was not sure big data was going to be great for everything. Inspired by the story of Donor Sibling Registry founders Wendy Kramer and her son Ryan, who'd managed to track down Ryan's anonymous donor father using Y-DNA testing and genetic genealogy all the way back in 2005, when Ryan was just fifteen years old, Erlich and colleagues had been working on a paper showing that they could fully identify men who'd anonymously donated DNA samples for research. They did this by taking advantage of the same correlation between surnames and Y-DNA that Alice had been hoping to exploit when she tested her brother Bill, and that genealogy researchers had been tracking in surname projects within public databases. Combining the power of those databases with the Internet, genealogical websites, and the limited "nonidentifying" information available about five research genomes, Erlich, his student Melissa Gymrek and their colleagues were ultimately able to identify not only those five men but many of their family members as well, altogether close to fifty people.

Erlich says there was some perception among his colleagues "that this is just fear-mongering," that he was pointing out a privacy loophole that only "boring academics" would bother to exploit. And at the same time, there was concern that his work would discourage the sharing of genetic data and make scientific research harder. "Some senior people told me that this might destroy my career," Erlich says. He was looking to move to an academic appointment once his fellowship was over, and this made him nervous, but he believed that what he was doing was important—that by alerting everyone to vulnerabilities, he might prompt solutions that would make things safer. At the very least, people participating in studies should know what they were getting into.

He submitted his paper, and it was accepted by the journal *Science*. On the day it came out, he told me, he went out to a celebration dinner with his wife and had to force himself to eat.

* * *

In April 2018, a major news event demonstrated the power of genetic genealogy. Investigators in California, matching crime scene DNA against the genetic profiles of ordinary people in the free, publicly-accessible genealogy database GEDmatch, uncovered the suspected identity of the notorious Golden State Killer and arrested him. Thought to be responsible for at least thirteen murders and fifty rapes in California in the 1970s and '80s, the Golden State Killer became known for his brutality and sadism. "As his victims lay terrified," the *New York Times* reported, "he would pause for a snack of crackers after raping them. He placed a teacup and saucer on the bodies of some of his victims and threatened them with murder if he heard the ceramic rattle." Victims' descriptions of him—a white man with blond hair—generated multiple sketches, while his crimes inspired terrifying nicknames: the East Area Rapist, the Original Night Stalker, the Visalia

Ransacker, the Diamond Knot Killer. Yet for decades, the culprit could not be found.

Police announced that the accused was Joseph James DeAngelo, a seventy-two-year-old former police officer living in a suburb of Sacramento. The notion that the Golden State Killer might be a cop had come up over the years, but according to the Sacramento district attorney, DeAngelo's name never had. "We weren't even close," Paul Holes, previously a cold case investigator with the Contra Costa County District Attorney's Office, told an interviewer.

Some months later, after hesitating over concerns for her safety, the genetic genealogist who'd assisted police stepped forward. She was Barbara Rae-Venter, a retired patent attorney with a PhD in biology who was trained by and then volunteered with DNAAdoption.org, the same organization Alice Collins Plebuch had turned to in order to hack her genes. Rae-Venter had not meant to be solving crimes—"I'm retired; I'm supposed to be working on my own family history," she explained during a lecture in Dublin, Ireland—and had essentially fallen into this work. Some years earlier, at the request of a detective, Rae-Venter had used segment triangulation (filtered through Alice's brother's iPad app, as it happens) in an enormously complex and time-consuming case to uncover the true identity of a woman who had been kidnapped, abused, and abandoned as a small child. Rae-Venter then helped determine the identity of the woman's deceased abductor, who police believe was a serial killer responsible for a number of deaths. Holes, who was nearing retirement, heard about Rae-Venter's work on those cases and contacted her to see if she might be able to help with the Golden State Killer, a case he'd been working on since the 1990s. Rae-Venter got on board and began to learn the details of her newest case. She thought Holes's criminal, whoever he was, made her previous serial killer "look like a choirboy."

Rae-Venter, Holes, and other law enforcement investigators got to work. They had the killer's crime scene DNA, which was uploaded as a profile to GEDmatch. Rae-Venter ran an ethnicity analysis of the crime scene DNA in GEDmatch and learned that the suspect had a fair amount of Southern European ethnicity. Through tools at that site and at Promethease, she got predictions that the suspect was likely bald and that his eyes were blue. She began the painstaking work of segment triangulation, teaching the investigators her techniques by phone and email. Because GEDmatch is much smaller than the consumer databases of 23andMe and AncestryDNA, the cousin matches for the crime scene DNA were quite distant, making the job challenging. But the team built trees for these matches and was able to identify a MRCA, or most recent common ancestor, and then to trace forward in time to candidates descended from that ancestor, men of the right age to have committed the crimes who had lived in California at the right time. Holes told me investigators were especially interested in five of these men. A relative of one of these candidates were approached, Holes said, and agreed to do "targeted" testing to help narrow down the pool, and this helped investigators home in on Joseph James DeAngelo. Holes told me the more he learned about the man—his physical appearance, his past behavior and associations, the areas where he was living during certain years—the more suspicious DeAngelo looked. (As it happened, Rae-Venter points out, DeAngelo was also the only one of the candidates with blue eyes. Plus, he had a receding hairline and ancestry from Southern Europe—from Italy, specifically.)

But that was not enough for an arrest. As in the many cases like this since that time, police then followed the suspect until he discarded his DNA in a public place—in DeAngelo's case, in a tissue in a trash can—and matched it to crime scene DNA. DeAngelo was arrested three weeks after Holes retired. Rae-Venter told me many of the serial killer's victims had continued to live in fear of his return; he's believed to have terrorized some by phone years after he brutalized them. When she watched the news of DeAngelo's arrest, Rae-Venter told me, she was struck by the thought that genetic genealogy meant not only would the alleged culprit see justice, but his victims might finally have some peace of mind. "They don't have to wonder that he's going to come through the door," she said.

If the Golden State Killer case was a wake-up call for many Americans about the unforeseen applications of genetic genealogy, there were

a number of people who were not surprised: members of law enforcement who'd already dabbled in trying to solve case through public databases, and certain people in the field of genetic genealogy, including
CeCe Moore, who'd known this kind of approach was possible for years
and had been thinking through the ethics of it. "The law has been lagging behind technology for a very long time," says Moore, who remembers speaking at the International Symposium on Human Identification
several years before the Golden State Killer case was solved, and walking
her audience through what she could do as a search angel. "They were
dumbstruck. They couldn't believe what we could do."

And then there was Yaniv Erlich, who had published that paper
in *Science* just five years earlier, when the experiment of using public genetic genealogy databases to identify anonymous research subjects struck some of his colleagues as unwise. In the end, that paper
didn't damage Erlich's career. The *New York Times* ran a story on his
"surprising" discovery, and he was invited to give a lecture and to give
testimony. He became a professor of computer science and computational biology at Columbia University, and later took a leave of absence
to become the chief science officer at MyHeritage. In addition to his
work on other things—including the construction of an astonishing
thirteen-million-person family tree offering insights into migration
patterns and genetic longevity—Erlich continued to work on genetic privacy and, in 2014, published a paper in which he surveyed all the ways
that privacy could be breached, warning about a then-obscure hobbyist
website called GEDmatch. At the time, he allowed, the threat was mostly
theoretical, but the world of genetic genealogy moves so fast that it took
only four short years to get from theoretical to arresting a man in one
of the most notorious cases in California history.

To understand why police were so excited to have access to GED-
match, it helps to understand how they usually use DNA to solve crimes.
For many years, forensic DNA testing by law enforcement has relied on a
different kind of technology than the one companies like AncestryDNA
use. Instead of looking at SNPs, police rely on STR testing—"the number of times certain short, known sequences repeat at various locations

on the genome," explains NYU law professor Erin Murphy, author of *Inside the Cell: The Dark Side of Forensic DNA*. The genetic markers law enforcement uses usually "reveal little to nothing about a person's medical or clinical history"—in contrast to the many SNPs employed by genetic genealogy companies, which are typically not used in a criminal justice context "in part due to these privacy concerns."

Police search for DNA matches by uploading the DNA profiles from crime scenes to local, state, and national databases that are known colloquially under the name CODIS. Looking at thirteen to twenty of these genetic markers known as STRs, the system compares crime scene DNA profiles to the millions of profiles of offenders, arrestees, and others already in the system. If the suspect is not already in the system—if he or she is not someone already known to law enforcement, in other words—police may be out of luck as far as this technology can help them. That's because "long-range" familial searching—trying to piece together the identity of someone who left their DNA at a crime scene by using relatives—is very difficult to do using the DNA profiles in CODIS, which are helpful only for finding the closest of relatives, like siblings, a parent, or a child. And besides which, not all states permit or practice familial searching, and those that do may impose restrictions.

To be sure, police are able to use CODIS databases to assist in solving crimes—hundreds of thousands of them, according to FBI statistics. But there has long been interest in taking advantage of the public databases filled with the research of genealogical hobbyists. Genetic genealogist Colleen Fitzpatrick consulted on the Golden State Killer case for California's Irvine Police Department as far back as 2011, using Y-DNA techniques to try to suss out the identity of the perpetrator. Reporting by several outlets shows that in 2017, investigators attempted to find the identity of the Golden State Killer by tracing DNA through a small, now-defunct public genealogy database for Y-DNA called Ysearch.org, run by FamilyTreeDNA. Law enforcement was able to get a subpoena to force FamilyTreeDNA to turn over some information about the account—"I explained to them why it wasn't going to help them, because the data they had was too puny," company president Bennett

Greenspan told me—and that led to the swabbing of an elderly man in a nursing home. (The man turned out to be an extremely distant relation to the Golden State Killer, Paul Holes told me.) But, as Rae-Venter pointed out to me, relying on Y-DNA alone often leads to mistakes and misidentifications. Neither that approach nor searches in CODIS had anywhere near the power and reliability of harnessing autosomal DNA through GEDmatch.

"We've had the Golden State Killer's DNA up in CODIS since 2001: no hits," Holes told a news station afterward. "We've done familial searches here in California every single year in which it's been legal: no hit. We've done international searches. We did everything without success. Law enforcement has been chasing this guy for forty-four years. . . . Once we started down this genetic genealogy road, it took us four months."

Using GEDmatch meant law enforcement didn't have to worry about whether or not a suspect or his brother happened to be in CODIS. So long as there were cousins, that could be enough. What's striking, population geneticists Graham Coop and Michael "Doc" Edge write, is that because of the richness of the DNA genotyping approach used by ancestry companies, GEDmatch's roughly one million people was vastly more helpful to investigators than the thirteen million DNA profiles then in CODIS.

* * *

At the time Rae-Venter and investigators used GEDmatch to unravel the alleged identity of the Golden State Killer, few of the hobbyists who'd placed their genetic information into the database could have anticipated that it would be used to solve a crime. For years, genealogists like Alice Collins Plebuch who'd had their saliva or cheek cells processed by consumer DNA testing companies had been downloading their raw data files and uploading them to this third-party site.

Founded in 2010 by a genealogist and retired businessman named Curtis Rogers and an electrical engineer named John Olson, GEDmatch is free and run by volunteers, though it charges a small amount for some advanced tools. A small private company developed in the spirit of open sharing and research, the site's policy was loose in 2018, warning users of the possibility that it could theoretically be used for purposes other than genealogy. And while it was much smaller than the databases of DNA testing companies like 23andMe and AncestryDNA, the appeal of GEDmatch to investigators was that it was accessible to them: 23andMe and AncestryDNA don't permit uploads of DNA processed elsewhere, and they, along with MyHeritage DNA, say they don't permit law enforcement to access consumers' genetic information unless they're required by law to comply with things like subpoenas and warrants.

Uploading crime scene data to GEDmatch required investigators to create a fake profile on the site. "An undercover profile," corrected Holes when a journalist asked about the ethics of this. "There's plenty of case law that allows law enforcement to pursue undercover activities to effect an investigation. So this is something that is lawful." Indeed, Erin Murphy told me, the only oversight of police using a genealogy database this way would come from the database itself.

As for that database, GEDmatch co-founder Curtis Rogers had not wanted his site used for criminal investigations. He later said that when it was used to break open the Golden State Killer case, it felt at first like a violation. But he was surprised by the reaction GEDmatch got: In emails, people were largely supportive and grateful. He came to feel the use of the platform in this way was a good thing. He and Olson alerted users and changed the site's policies to say that law enforcement could upload DNA for the sake of solving murders and sexual assaults and for identifying remains. Law enforcement officers across the country became aware that they had a powerful new tool that could radically improve their ability to solve some of the worst crimes.

Within weeks, a DNA technology company outside Washington, DC, called Parabon NanoLabs made an announcement. Parabon, which had previously offered law enforcement agencies a service that creates a composite sketch of a suspect or victim based on DNA, now had an investigative genetic genealogy unit helmed by CeCe Moore. Less than a month after DeAngelo's arrest, Parabon had already uploaded genetic information from about a hundred crime scenes into GEDmatch, and Moore had begun to unravel them, at a typical cost to police departments of about five thousand dollars per case. I spoke with CeCe Moore that summer, and she told me she'd long wanted to use DNA in this way, but for years had turned down requests from detectives, coroners, and district attorneys. At first, she declined in part because GEDmatch's database was too small to be useful. But more broadly, she had ethical and legal concerns. She couldn't do this kind of work behind the backs of the genealogy community, she felt, especially since she was responsible for encouraging many people to test. So, for a long time, she sat on her hands, and *not* doing this work was its own kind of burden. If a perpetrator was still out there actively committing crimes, was her inaction resulting in more victims?

The Golden State Killer case, and the publicity around it, had blown Moore's concerns away. "I couldn't have asked for a better advertising plan to get the word out," she said. Anyone who didn't want their data to be used that way could take it down. Besides which, the case seemed to make clear that she was legally protected if she did this kind of work, and, as she gauged the reactions of bioethicists who weighed in, she came to believe her gut instinct was vindicated.

"I just think it's the right thing to do," Moore told me. She frames investigative genetic genealogy as akin to the work she's done with adoptees and the donor-conceived. "My work is about family and trying to bring closure and answers and peace, and it's exactly the same thing—it's just a different avenue to do that." She told me she was still consulting to *Finding Your Roots*, but had scaled back on her search angel work to make room for a second full-time job at Parabon. Sixteen-hour shifts had turned into twenty-four-hour shifts. "I've had cases where I just could not walk away," she said.

By May 2019, a year after Parabon had launched this service, it announced it had helped solve fifty-five cases. Many of these were cold cases involving crimes that took place well before commercial DNA databases and genetic genealogy existed, not to mention the Internet; the oldest had been committed more than fifty years earlier. Working from her couch, Moore was able to tease answers from the old crime scene DNA and give the names of possible suspects to police as tips for them to confirm. Many of the cases were jaw-dropping in their horror. They included a man in his fifties arrested for raping and strangling an eight-year-old Indiana girl in 1988 (he pled guilty and was sentenced to eighty years), and a man in his thirties arrested for the rape just a few months earlier of a seventy-nine-year-old woman.

Meanwhile, Barbara Rae-Venter has also been working cases, offering a sliding scale and even working pro bono when local agencies' budgets are tight. The FBI flew her to Texas to teach federal agents and police her methods. And Colleen Fitzpatrick has been helping with murders and unidentified remains, adding autosomal DNA techniques to her old methods involving Y-DNA. In the months and years after DeAngelo's arrest, the faces of long-ago victims were revived again in news stories: mugging for the camera or smiling from school photos. Most were women and girls, and the crimes perpetrated against them often involved sexual assault. Now, these gendered crimes were being solved by female citizen-scientists, who were pointing police toward men who, if they were guilty, had lived for decades without any repercussions for the most atrocious crimes. It prompted a kind of reckoning, a sense for some that this is the way the world is supposed to work and that genetic genealogy was making it work that way.

Sometimes, Moore says, she sweats while looking for a culprit, cries upon finding him. These discoveries can be heavy things because once she pieces her way to a likely suspect, there is a period of time when she is perhaps the only person in the world who knows his identity, other than the criminal himself. In June 2019, one of Moore's solves became the first case involving genetic genealogy to go to trial and lead to a conviction after defense attorneys didn't challenge Moore's techniques, agreeing to treat her work as "just a scientific tip," as she puts it. "That's really good news for us." After a jury pronounced William Earl Talbott II guilty of murdering a young Canadian couple in Washington State in

1987, a local investigator said genetic genealogy was "an amazing tool, and we'll be using it again."

But a number of privacy experts are uneasy about the way that Moore, Rae-Venter, and others have been helping police make arrests. In a paper published in *Science*, University of Baltimore law professor Natalie Ram and others argue that while "there are few legal roadblocks to police use of genetic databases," this may nonetheless violate the spirit of American law. "The Fourth Amendment's protection against warrantless searches and seizures generally does not apply to material or data voluntarily shared with a third party"—in this case, a platform like GEDmatch. Yet part of the Fourth Amendment's civil rights protections involves "a commitment to protecting privacy or freedom from government surveillance," Ram and her colleagues wrote. "Police cannot search a house without suspecting a specific individual of particular acts—even if doing so would enable the police to solve many more crimes." In other words, solving crimes is not the only thing American law values; our right to privacy isn't—or shouldn't be—subservient to this interest.

"The fundamental order that's set by our Constitution and by the protections in the Fourth Amendment and elsewhere . . . is one that says people should be left alone by the government in its policing capacity unless the police have a suspicion, a reason to think you've committed a crime," says NYU's Erin Murphy. "Something as simple as your name can't be forced from you by police without suspicion." DNA databases, Murphy argues, are part of a larger trend in which new technologies involving big data and surveillance upend that presumption. "DNA, when used to find suspects or to put people under suspicion who otherwise would not have reason to be under suspicion, is an inversion of how we actually do things." And the shared nature of genetic data means that "this idea of consent or this idea of third-party disclosure doesn't really work," she continues. After all, when a man uploads his genetic data to GEDmatch, can his sister be considered to consent to this? Or, as search angel Laurie Pratt put it after pulling her genetic data from GEDmatch: "Do I have the right to be your genetic informant?"

But Paul Holes argues that these new techniques can be less invasive than a traditional investigative strategy relying on "the thousands of
tips that have come in because somebody looked like a composite drawn
back in the 1970s." With previous methods, he told the *New York Times*
podcast *The Daily*, "we gathered DNA from hundreds and hundreds of
people because of these tips and because of circumstantial evidence
within the case file." Besides, he pointed out, "we aren't accessing anybody's actual genetic information. We are just seeing how closely or
how distantly related some people are to who our offender might be."
Rae-Venter told me she thought it illogical for people to object to her
using the same techniques for crime-solving that she'd long employed
in her search angel work. "Nobody's jumping up and down about anybody using the database for identifying people who were sperm donors,"
she said, and plenty of those donors were actually promised anonymity,
unlike criminals leaving their semen in the act of committing rape.
Most Americans appear to agree with Holes and Rae-Venter; one study
of about 1,500 people surveyed online in the United States found that a
sizable majority supported police searching of genetic genealogical databases to identify perpetrators of violent crimes, perpetrators of crimes
against children, and missing persons.

What's needed, observers on both sides of the debate told me, is
oversight. Thomas Murray, the Hastings Center president emeritus,
told me he supported the use of genetic genealogy "to prevent the most
extreme sorts of harms," so long as there are "good guardrails" in place
to make sure police don't use these tactics for lesser crimes. A number
of scholars believe there need to be legislative fixes that at the very least
delineate when and why police are allowed to utilize genetic genealogy
databases, whereas a state legislator in Maryland proposed a law ban-
ning the use of genealogical databases for crime-solving altogether.
And yet, not long after I spoke to Murray, news outlets reported on
the use of genetic genealogy leading to the arrest of a South Dakota
woman "charged with murder for allegedly leaving her newborn in a
ditch thirty-eight years ago." Several genetic genealogists expressed
concern about a slippery slope, about the kinds of crime this powerful

new technology was being used for, and about the implications of such cases for women's reproductive and bodily autonomy.

Another concern among genetic genealogists, who know the perils of making mistakes in their work, is that no one is making sure these new crime-solving techniques are executed properly. "I trust somebody like CeCe—she's really good—but we don't have any protocols for genealogists doing this type of work and what the best practice is, and the police themselves don't know who to use," Debbie Kennett told me. Kennett has heard of cases in which adoptees were reunited with the wrong parents because of mistakes in interpreting data, and she worries about questionable genetic genealogy techniques leading investigators astray.

In fact, this has happened, most famously in 2014, when, according to a story in the *New Orleans Advocate*, Idaho investigators trying to solve the brutal 1996 murder of a woman named Angie Dodge became interested in a New Orleans filmmaker in his thirties. His name was Michael Usry, and he became a suspect because Y-DNA from the crime scene was partially matched to the DNA of his father, who years before had submitted a sample to the Sorenson Molecular Genealogy Foundation. Ancestry, which had acquired Sorenson's assets, received a court order to reveal the father's name, and investigators traced their way to the son. They got a search warrant ordering Usry to provide his DNA, and after about a month it turned out that he was not a match. Shortly afterward, Ancestry made the Sorenson database private to protect customers' privacy, a loss Moore mourned as akin to burning a library. "Civil rights are being taken from us at an alarming rate," Michael Usry told the *Los Angeles Times* a few years later.

And yet, autosomal DNA tracing techniques in the hands of a skilled and experienced genetic genealogist are much better than the techniques employed in the Usry case, which Rae-Venter told me shouldn't even be considered investigative genetic genealogy. And besides, advocates say, this is why police confirm the tips that Rae-Venter and Moore give them. (Other apparent problems in the Angie Dodge investigation underscore the fact that DNA is not alone in its potential for police

misuse. In 2017, another man who'd been implicated by the investigation was freed and would eventually be exonerated following two
decades in prison after giving what he now says was a coerced confes-
sion to aiding and abetting in Dodge's rape and murder.)

In 2019, as if in vindication of investigative genetic genealogy's
promise, police in Idaho announced the arrest of another man for
Dodge's murder. This time, they'd homed in on their eventual suspect
using GEDmatch, through the work of CeCe Moore.

* * *

In late 2018, Yaniv Erlich published another paper. This one contained
the revelation that, from the genomic data of 1.28 million people, it was
possible to find a third cousin or closer for 60 percent of Americans of
European descent, the primary population engaging in genetic testing. This meant that all those Americans, the vast majority of whom
had never spit into a tube or swabbed their cheeks, could potentially be
identified with additional research. Consent truly didn't matter anymore. Most Americans would soon be consented for, whether or not
they knew it.

This time, Erlich told me, he didn't get pushback on his findings.
So much had changed in five years that his colleagues didn't question
his premise that genetic privacy was profoundly compromised. They
didn't suggest that by showing what was possible, he was acting unethi-
cally. "I think this is part of the maturation of our society that we now
are well aware that big data can be abused," he says.

The statistical implications of Erlich's paper were that once a database contained about three million people of European descent, 99
percent of the greater population of these Americans would be potentially identifiable, at which time the United States would essentially
have what legal scholar Natalie Ram describes as a de-facto national
DNA database. At the time Erlich pointed this out, it seemed the
prediction wouldn't be a reality for some time, till the database of a

third-party website like GEDmatch got big enough. But a few months after his paper was published, news broke that almost overnight, the population that law enforcement had access to had doubled, to about two million.

BuzzFeed News reported that FamilyTreeDNA was allowing the FBI to search its database for relative matches of crime scene DNA to identify perpetrators of certain violent crimes, as well as the remains of John and Jane Does. It was early 2019, and the company told *BuzzFeed* that it had been processing crime scene samples for the agency in its lab and uploading them "on a case-by-case basis" for several months by that point. A number of genetic genealogists, who had regarded Bennett Greenspan's company as one created by a genealogist for genealogists, said they felt betrayed. Judy Russell, a former lawyer and newspaper reporter known for her blog *The Legal Genealogist*, pointed out that FamilyTreeDNA was giving law enforcement access, in many cases, to genetic relatives' full names, email addresses, and family tree information.

In the days and weeks that followed, FamilyTreeDNA unleashed a blizzard of statements and policies, eventually announcing that consumers in the United States could opt out of having their DNA used for law enforcement purposes if they wanted to. And Greenspan offered a confusing explanation as to what in his company's terms of service had permitted him to give law enforcement access to his database—even as Russell told me the contractual language appeared to say the opposite. I'd spent months researching privacy policies and clickwrap contracts by then, and this seemed a striking example of the problems around informed consent in the context of genetic data. If even Russell, with a law degree, couldn't make heads or tails of a company's policies, what hope was there for the average consumer?

It turned out the process of becoming involved in genetic crime-fighting had been rocky not just for Greenspan's customers but also for Greenspan himself. When I spoke with him a few months after the *BuzzFeed* story broke, he told me he'd had to grow a new skin over the past few months and that he'd given his customer service representatives "combat pay" for the days following the article. And he explained that while he did not have a philosophical objection to his database being used to solve violent crimes, he hadn't set out to make this stand. He said his hand had been forced by the fact that, unlike 23andMe and Ancestry, FamilyTreeDNA permits people who've had their DNA processed by other companies to upload their data to its site, which meant law enforcement could theoretically upload crime scene data so long as it was compatible. And indeed, he said, at one point the company discovered that someone was attempting to upload files in a strange format. Greenspan told me he believed this someone was the FBI.

Greenspan told me he had a moment of reckoning. By then he'd received three subpoenas for information in the main FamilyTreeDNA database or in the now-defunct public database Ysearch.org. He was fielding phone calls from an FBI attorney. Law enforcement was trying more than one approach to access his main database, and he came to feel its presence there was inevitable. The best approach, he decided, was to work with law enforcement and regulate the situation by subjecting cases to approval. By the time we spoke, Greenspan had decided to lean into the controversy, convinced he was on the right side of history. His company had released an advertisement starring Ed Smart, father of kidnapping survivor Elizabeth Smart, encouraging people to upload their genetic data to FamilyTreeDNA for free to help victims and their families. He quibbled with me over whether it was fair to say he'd permitted law enforcement "access" to his database; his preferred description of what he was doing was "not obstructing justice." But he was cagey about certain details and the account he gave me differed from one he'd given a publication called *Forensic Magazine*. Months later, in an interview with the *Wall Street Journal*, Greenspan offered a third version of events. He now said his company's relationship with the FBI began when he willingly cooperated on two criminal cases because it was the right thing to do. (Meanwhile, a company spokeswoman walked back Greenspan's claim that the suspicious upload he'd told me about was actually the FBI, saying he'd been mistaken.)

Yet the use of genetic information to solve crimes, however worthy, has to be balanced with legitimate concerns on the side of protecting privacy, Erlich told me. The problem is one of trust. "Genetic information is essential if you want to advance precision medicine," he said. "We have families, some of them I know personally, that genetics was able to help, to identify the cause of serious conditions." But what if people were to stop trusting researchers to keep their information safe? "If we cannot recruit tens of thousands, maybe hundreds of thousands of people for these studies, we cannot help these people," Erlich said. "We cannot use the power of the genetic revolution to empower our understanding. And that's a huge missed opportunity."

This erosion of trust may already be happening. In early 2019, 23andMe co-founder and CEO Anne Wojcicki said the market for her product had slowed, speculating that this might have to do with broader privacy concerns stemming from things like the Golden State Killer case and Facebook's dicey data-collection practices. Some observers have suggested that the lack of privacy protections around genetic data may disproportionately affect minorities, and I heard a number of accounts of African Americans quitting GEDmatch after it became clear that police were using it.

And, of course, concerns about how we protect our genetic information extend beyond the context of recreational testing. Some experts worry about the increasing use by police of Rapid DNA machines, which can process samples in ninety minutes, and which they fear poses the possibility of errors and privacy violations. There are serious concerns about DNA testing on migrants. And all of this is taking place within the broader context of what legal scholar Andelka Phillips calls "ever-increasing monitoring, where we are all subject to complex data mining and profiling." The problem, Phillips says, is that it's difficult to imagine how data from direct-to-consumer testing may in the future be linked with other data and used in ways we can't anticipate.

It might, in theory, be used by some future totalitarian government to discriminate against its citizens. This may seem implausible, except that the day I spoke with Erlich—and asked him if he could imagine a future world in which our ancestral backgrounds could be used to hurt us, and he said he didn't need to project into the future because "I can go to the history"—on that same day, the *New York Times* ran a story about how Chinese authorities were using DNA as part of a "campaign of surveillance and oppression" against the country's ethnic minority Uighurs.

Journalist Kristen V. Brown has thought a lot about big data and unintended consequences. In 2015, she covered the hack of the website Ashley Madison, an online dating service geared at people who wanted to cheat on their partners. "Life is short. Have an affair," the site's slogan went. Brown wanted to know how the hack and subsequent exposure of the site's users' personal information had changed their lives, so she interviewed more than one hundred people impacted by it and learned about divorces, blackmail, and suicides. It's easy to assume that if you don't cheat on your spouse, you don't need to worry about this sort of thing, and that if you do cheat, well, you deserve what you get. But Brown thought the incident had broader implications. She was fascinated by a concept from Georgetown Law professor Paul Ohm that we are all at the mercy of the massive troves of data that businesses collect and keep on us, and that somewhere amid all that information, every one of us has a devastating secret. Ohm called this eventual, interconnected treasure trove of information the "database of ruin," and he urged in a 2012 *Harvard Business Review* article, *Please don't build this.*

This idea stuck with Brown, and when she started covering consumer DNA testing, she saw how this new technology fit the paradigm. "I was like, 'Oh my god, our genomes have now become databases of ruin.' It's just another piece of data that can be incriminating," she says. "I think we're at the beginning of living in a time where you can't really keep secrets anymore. And how does that change how we go about the world and our daily lives?"

How does it? We're such bad prognosticators. Historian Melvin Kranzberg once wrote that technology is neither good nor bad, nor is it neutral, by which he meant technologies play out in vastly different ways depending on the context, and that we often lack the ability to anticipate how they will change our lives. "He realized that the impact of a technology depends on its geographic and cultural context, which means it is often good and bad—at the same time," columnist Christopher Mims writes in the *Wall Street Journal*. The pesticide DDT, heralded for knocking out malaria and then reviled as an environmental threat, was the example Kranzberg used; Mims invoked Facebook. Consumer DNA testing seems a pretty good example, too.

Just over a year after Joseph James DeAngelo was arrested, the rush to that de facto national database abruptly slowed. After a seventy-one-year-old woman in Utah was attacked and choked into unconsciousness as she was practicing the organ in her church, GEDmatch co-founder Curtis Rogers made an exception to the site's policy that law enforcement access the database only for murders and sexual assaults, and allowed police in to help solve a case he described as "as close to a homicide as you can get." In the ensuing controversy over the site unilaterally making an exception to its terms of service, GEDmatch decided to expand the definition of violent crimes that law enforcement matching could be used for—and to automatically opt all its users' information *out* of being available to law enforcement. If people wanted their genetic information used in this way, Rogers decided, they needed to proactively choose it. Overnight, the database of people that CeCe Moore and Barbara Rae-Venter could access to help police narrow in on suspects went from about a million to zero. Moore called it "a setback for justice."

When I spoke with Rae-Venter a few weeks after GEDmatch's decision, she told me she was hopeful that a growing opt-in movement would eventually make the database a destination for law enforcement matching again. By then tens of thousands of people had logged into GEDmatch to request that their genetic information be used to solve crimes. In the meantime, she said, she was still able to use FamilyTreeDNA for investigations. Some months later, FamilyTreeDNA announced a new investigative genetic genealogy unit to rival Parabon's, headed by Rae-Venter. And then the *New York Times* reported that a Florida detective had obtained a warrant giving him permission to search *all* of GEDmatch's database, including data from the majority of users who hadn't consented to being involved in criminal investigations. The judge's decision to grant this order was a game-changer, experts told the paper; it would likely embolden other agencies to seek similar warrants for huge companies like Ancestry and 23andMe, potentially "turning all genetic databases into law enforcement databases."

DNA revelations mean that an alleged sadistic serial murderer named Joseph James DeAngelo is behind bars, but they also mean that you could be implicated in the arrest of a relative. It means that an adoptee can find her parents, and it also means that a family can find out that decades before, a man cheated on his wife and produced a child—which could be good for that child and bad for that wife and a mixed blessing for that man. Perhaps consumer genomics means we are all on guard waiting for the other shoe to drop, or perhaps it means that we are all forced to be more honest with each other. One of the central conundrums of spitting into a tube is the way one person's rights so often collide with another's after the tube is sealed and sent in.

15

LATE NIGHT

By early 2015, two and a half years had passed since Alice had first spit into a tube. She and Gerry had folded a platoon of people into their effort—from a rabbi in New York to a professor in North Carolina, from genetic genealogists in California and Uruguay to distant cousins in Massachusetts and Minnesota. But in the end, it was not any one of these people who provided the crucial bit of evidence the Collins sisters needed. The missing link came in the form of a genetic stranger who was not looking to solve a mystery. This stranger did not know, when she got her own results, that they were much of anything to ponder or that they would profoundly alter her own family as well as another.

Alice and Gerry were monitoring quite a number of DNA kits by then. Mainly, they checked on their own results and those of their brothers, looking to see if any new DNA cousins had shown up who might help them unravel family relationships and solve their case. Alice was also managing the account for her first cousin Pete Nolan, the son of her father's beloved sister, Kitty, but she would check on it only occasionally, as it rarely turned up any new close cousins. The significance of Pete was that he *wasn't* genetically related to the Collinses—as Alice and Gerry had learned after testing him at 23andMe. Pete's mom, Kitty, had been Jim's sister, but they were not genetic siblings, and it was this crucial bit of evidence that allowed them to understand that Pete was the biological grandchild of John Josef and Katie Collins, whereas they and their siblings were not. This, in turn, had allowed them to discard several of the theories they'd come up with to explain their unexpected ethnicity, and to conclude that their father must have been confused with another child, most likely when he was born at Fordham Hospital. But they had no conclusive proof of this theory and no inkling who that other child might be. If their father had been taken from another family, what family was it? So far, none of their efforts to unravel the truth had worked.

One day in January, Alice decided to send Pete an email updating him on what was happening or, more precisely, not happening, with her search. She logged onto his 23andMe profile, not expecting anything interesting.

But something interesting had appeared. A new close cousin had shown up at the top of Pete's list of relatives. The significance of this anonymous woman and her relationship to Pete were not clear, but Alice was intrigued. Since Jim and Kitty were not biological siblings, Kitty must have had a genetic brother she never met—the baby boy born to her parents, Katie and John Josef Collins, and switched with Jim. And that baby boy was presumably raised by the *other* family, the one Jim should have gone home to, and might have grown up and had children of his own—and one of his descendants might be interested in genealogy and might have sent her saliva into 23andMe. And this person might be showing up as Pete's closest cousin at this very moment. Alice, staring at her iPad in the family room, began to shake. She had a prescient feeling.

Alice wrote the woman through 23andMe's internal messaging system and asked if she would be willing to compare genomes with Pete Nolan. She explained that she was Pete's cousin and that she was trying to solve "a hundred-year-old mystery concerning my father." Then she waited at her iPad, worried that the stranger wouldn't reply, as so often happened with these requests. But a short time later, the stranger did reply, mutely signing into her account and allowing for genome sharing. Alice studied all the places where the stranger's DNA overlapped with Pete's and concluded that the stranger was probably a first cousin once removed based on the centimorgans—an extremely good match. But was the stranger related to Pete in a way that would be helpful to Alice? She wrote again and asked the woman what she knew about her lineage.

The woman on the other side of the exchange studied the screen. Her name was Jessica Benson, and she was that Raleigh, North Carolina, stay-at-home mother whose ethnicity results through her paternal side hadn't been what she'd expected, but who had not been transformed into a seeker by the discrepancy. Jessica's dad and his parents were no longer alive, so she'd gone to her dad's sister with her unexpected ethnicity estimate, but her aunt couldn't shed any light on it, and Jessica had thrown up her hands. Yet one lesson of the consumer genomics age is that even when you don't go looking, the truth can come looking for you. In this case, the truth took the shape of Alice.

"I was actually expecting to be much more Ashkenazi than I am," Jessica Benson wrote back. The weird thing was, 23andMe was telling her "that I am actually Irish, which I had not expected at all."

Alice got chills. She wrote again. Her father had been born at Fordham Hospital on September 23, 1913. Had anyone in Jessica's direct ancestral line been born on that date in that place? Jessica didn't know. She didn't even know where Fordham Hospital was. She googled it and learned it had been located in the Bronx, where her dad's family had lived. She pulled up the Social Security Death Index and wrote back that it appeared her grandfather, Phillip Benson, had been born right around that date.

Alice began to cry. She went downstairs and told her husband, Bruce, she thought she'd found something big. She started looking through the pages of the New York City Birth Index of 1913, trying to figure out how she'd missed the name Phillip Benson, if indeed such a baby had been born in the Bronx around September 23. If this child were real, he had to be in there.

It was looking increasingly likely that the birth certificate Alice had for "James Collins" was, in fact, not properly her father's birth certificate. It had become his when he'd been handed to the wrong mother and gone home with the wrong family. But when he was born, before capricious fate confused him, Alice's father had been "Phillip Benson," and Phillip Benson had been James Collins, and everything that followed had flowed from this single mistake. It was a wild thing to contemplate, and yet it was the simplest explanation given everything Alice knew.

It was nearly midnight when she finally found the right page among her 1913 baby names. In her thick white binder, there was one line that she had circled and put an asterisk next to. The name was blurry. The birth certificate number was completely illegible. The first name was Phillip, albeit misspelled, and the last name was something—Bamson or Bamason or Ban_son. During those months combing the index years earlier, Alice had struggled with this name and placed a turquoise sticker with a question mark next to it.

But now she knew: "Philip Bamson" was really Phillip Benson. She sent Jessica an email with a screenshot of the page: "I FOUND HIM!" Alice barely slept that night.

* * *

The next day, Alice emerged from sleeplessness and began to tell people about what had happened: her siblings, CeCe Moore, and her DNA cousins, who immediately started scouring the Internet looking for archival information on the family of Phillip Benson.

Jessica Benson began sending photos of her father's side of the family. Alice was in a sewing class when she got one of her emails and saw, for the first time, an image of a man named Sam Benson, who had been Phillip Benson's father—and who (it was now becoming clear) was Alice's genetic grandfather. She was stunned by how much Sam looked like one of her brothers. Meanwhile, Jessica put Alice in touch with her aunt Pam. Pam Benson is the daughter of the late Phillip Benson, which is to say she was like Alice on the other side of the looking-glass—a woman in the process of learning that her father was not who she thought, and neither was she.

In retrospect, the conversation between Alice Plebuch and Pam Benson has to have been one of the stranger iterations of a modern-day game of telephone. After Alice's long search for something else entirely—for her genetic cousins—she had found the daughter of the man her father had been switched with. Only, she didn't realize it; she assumed that Pam had married into the family. She had never intended to contact a direct descendant for fear of disrupting someone else's life as hers had been disrupted.

Meanwhile, Pam, who lives in Lawndale, California, was not really prepared for this conversation, if it's possible to prepare for such a thing. The news that her dad had been switched at birth with another child unfolded in a rapid and confusing fashion. Her first inkling that something was strange was that after her niece Jessica had taken a DNA test, Jessica had contacted her saying, "Auntie Pam, do you know we're Irish?" To which Pam had replied: "You're nuts. The test is wrong." Pam told me she went through "this whole spiel" explaining why Jessica's results were impossible—her father had always said his family were Russian Jews, and the genealogical paperwork confirmed it. Then, some time later, Jessica's mom contacted Pam again with a crazy story involving a woman named Alice, and could they give Pam's phone number to this woman?

And so Pam and Alice spoke, and Alice laid out her theory, and Pam's reaction was "You've got to be kidding me," and for some time Alice didn't realize she was talking, in essence, to herself. She was telling another woman that her father had been swapped in the hospital and originally destined for a different family, a different culture, a different religion, and a different life. When Alice put everything together, she felt terribly guilty about the way that first conversation had gone with Pam.

Pam ordered her own DNA test, but well before the results came back, she sensed in her bones that Alice was right. She saw a photograph of John Josef Collins, the man Alice had always known as her grandfather, and it was clear that he was, in fact, Pam's grandfather. He looked just like Pam's brothers. And it was becoming clear why Pam's own grandparents, who'd died before she was born, looked nothing like her father. Pam's grandfather Sam had been five foot two, and her grandmother Ida four foot nine. Their son Phillip had been six foot four. In a photograph, Sam barely cleared his son Phillip's shoulder, and they looked absolutely nothing alike. "I always wondered, 'How could Grandpa be so short and Dad be so tall?'" Pam told me. She'd asked her dad once for an explanation, and he'd said, "recessive genes."

Phillip Benson, center, with his father, Sam, at right, and a relative

Now, Pam was where Alice had been, only Alice had had years to digest the news of a disrupted identity and make sense of it. She had been an agent in her own discovery. Pam had had the news thrust upon her, and it was jarring. She was in her fifties by then, nearing retirement. How could she suddenly "go from this to this"? Pam asked me. "It's, like, nuts," she said, using her favorite word. There were times when Pam's daughter would find Pam crying. "What's wrong?" she'd ask her mom. What was wrong was—well, everything. How was Pam to reconcile her new identity when she'd lived so long with an old one? She was *of* her father, and now her father was someone else. Pam had not been raised Jewish—her mother, Phillip Benson's second wife, was a Lebanese American Catholic—but "that was what I knew my dad was," she says. Growing up, she celebrated Hanukkah and Christmas; her parents sometimes went to friends' houses for Passover Seders. To learn now that "my dad was really Catholic," and Irish—this jarred Pam. It was as if she had to let go of something on her father's behalf.

And she felt sad for him. She imagined her tall, blue-eyed father sensing all his life that he was out of place in his family, towering over his parents, looking not a whit like his sister. It was too crazy-making to have this knowledge and imagine Phillip in ignorance of it. Somehow, the revelations had "altered the course of time," Pam said, and made the past present. More than three decades after her dad's death, she spoke of calling questions up to him in heaven. She pondered how he'd made sense of his difference—if he'd questioned whether his parents were truly his parents or wondered whether his mother had cheated on his father.

Phillip Benson in his later years

And so, when Pam's daughter asked what was wrong, Pam would say, "I'm just wondering what my dad would have thought. Would it have answered questions for him?" Within that question was embedded another: Who was Phillip Benson, really? Who was Jim Collins? What makes us who we are? Blood? Family? Culture? At one point, Pam thought back on her father's character, the generosity he showed throughout his life, and concluded that this was innate. "The family that you're given can alter your environment and the way you're raised but it's not going to alter your soul," she said. "Your soul is given to you by a higher power."

Things might have unfolded differently if she'd made the discovery herself, Pam told me. It might have eased her initial sense of disbelief and befuddlement. "You have more control when it's you," she said. Indeed, "there was a split second when I wished that the DNA wouldn't have been done, and I would have gone on happily ever after." But then, as Alice sent old photos and began to fill Pam in on her father's family of origin, and told her she had a first cousin she'd never heard of—a man named Pete Nolan—the truth became fascinating to Pam.

"I believe that it is better to know," Pam told me. The truth answered so many questions. "I mean," she added, "I still say it's nuts."

* * *

The incontrovertible proof of her theory, Alice learned following her 23andMe breakthrough, lay in the DNA of a woman named Phylis Pullman. Everything pointed to Phylis being the biological cousin she'd never known.

When Alice's father was born—intended for the life of Phillip Benson, but transformed by an accident of fate into Jim Collins—he already had a sister, a toddler he never met. That girl, Miriam, grew up—well, sort of; she never quite reached five feet—and had a child named Phylis, who is also small in stature. Alice got in touch with Phylis's daughter and laid out what she allowed was an "absolutely bizarre" tale, and

Phylis's daughter reached out to Phylis and said, "Mom, I got the strangest email."

The first time Phylis and Alice spoke by phone, Alice was struck by the way Phylis used a phrase that had been one of Jim Collins's favorites. "Such is life," she said, and in those words was so much: a shrug and a laugh and a shaking of the head in wonder at the strange diversions of fate. Phylis sent away for a DNA test and began to share family stories. As Alice would learn, Phylis, now in her eighties, did not know her uncle Phillip well, because he'd moved to California when she was young. But as a little girl she perceived him to be "a giant" in contrast with her own petite family, even if no one around her appeared to make much of this. Yet others may have been struck by Phillip's differences. Phylis told Alice how, when Phillip was courting his first wife, her observant Jewish parents didn't believe he could possibly be a member of the tribe. "He had to bring his birth certificate," Phylis said. "Little did we know it wasn't *his* birth certificate."

Through stories and genealogical research, Alice began to fill in the story of her grandparents' lives. Like the Collinses, the Bensons were immigrants during the first decade of the twentieth century. They came over as young adults from the same part of the world as each other, part of a large wave of Eastern European Jews escaping poverty and discrimination who settled in large cities like New York. Sam's and Ida's documents generally listed "Russia" as their birthplaces, but Alice is a digger, and eventually she was able to figure out that they came from two towns in Lithuania with sizable Jewish populations. About thirty-five years after Sam and Ida left, the Jews in these towns were murdered during World War II German occupation. The places Sam and Ida emigrated from are within about an hour of Vilnius, the city that had been blanketed in yellow "ancestor pins" when Alice had mapped out her DNA cousins' ancestral villages, and now that pattern made sense. She'd known her people came from there; it just took finding Jessica Benson to figure it out, one accident of fate correcting another.

According to accounts, Ida was sweet and nurturing, whereas Sam could be taciturn and cold. When Phillip was born, the Bensons appear to have been living in a five-story apartment building that was just a ten-minute walk from the Collins family; Miriam Benson and Kitty Collins, both toddlers, could have played with each other at the big park just up the road. The Bensons then moved around before eventually settling in Brooklyn. Sam started in the position of "button hole maker," then became a machinist in a factory. Like his biological son Jim, Sam loved to play cards; unlike his son, he lost the family paycheck more than once, forcing his wife Ida to borrow money from family. Eventually, Sam ran a shop where he repaired, of all things, Singer sewing machines—another tiny accident of fate that parallels his granddaughter's love of high-tech sewing machinery. What was his shop like? Was it like Alice's office, crowded with machines and other things, always barely on the safe side of chaos? And what was his mind like? Could he, like his son, follow all the moves of an imaginary game of pinochle in his head? Did he have an eye for detail so that he could—as Alice did with computer programs—look at a sewing machine and rapidly pick out the "nit" that was making it break down?

Or is it facile to assume such a parallel? After all, Phillip Benson also took after Sam, the man he knew as his father. Like Sam, Phillip had a talent for taking things apart and putting them back together. When he was sixteen years old, he was listed on a census as a mechanic working in an automobile garage. Later, he helped run a company that built custom airplane parts.

When Phylis's DNA test results came back, it confirmed what she and Alice already knew: They were first cousins. When Pam's DNA test came back, it confirmed that she was first cousins with Pete Nolan, and she began to email with him. Alice and Pam, whose fathers had been switched in the hospital, began calling each other "swapcuz." And Alice and Pam and Phylis, making up for whole lifetimes, embarked on what I call "The Great Exchange," swapping visits, photographs, family stories, and memorabilia. Alice flew to visit Phylis in Florida, and the first time

Phillip Benson and his first wife, Esther Abolafia Benson,
sitting with their son Kenny. Behind them are the parents
Phillip grew up with, Ida and Sam Benson.

the first cousins sat across from each other on a couch, inadvertently holding the same pose, Phylis's daughter laughed and told them they looked like mirror images. They have the same lips, the same nose, the same eyes, and the same build. They were born with identical gaps in their front teeth.

Phylis told me she was thrilled to discover she had seven first cousins and all their spouses and children—she'd long considered herself rich in family, but now, her wealth had doubled. She told me that getting to know "the Collins-Bensons," which is what she calls Alice and her family, "has been the most exciting experience of my life." But the discovery was tinged with sadness that she'd gotten this news so late in life. When Phylis was sick, Gerry went to Florida and cooked her meal after meal, the way you do for family (including family you didn't know existed till you took a DNA test). "I've only got so much time with her," Gerry told me.

And, in early 2016, Alice flew south to California to visit her swap-cuz Pam, who, at five foot ten, takes after her father, Phillip, in height.

Alice's head barely clears her shoulder. The swapcousins were not related by blood or marriage but they were related in a way, by a mistake a century earlier, one that only DNA testing could have uncovered. "It was wonderful," Alice told me. "We essentially had experienced a lot of the same things, just from the other side." Pam felt moved to give Alice things that had belonged to her grandparents: the Bensons' naturalization papers and Sam Benson's Social Security card. Alice sent her swapcuz the 1918 war zone pass that had belonged to John Josef, the man she thought was her genetic grandfather but who was, in fact, Pam's.

It took Pam a while to get her father's birth certificate from New York State, in part because Phillip's last name had been misrecorded as "Bamson." But when it arrived, she and Alice could see for themselves exactly how one moment long ago in a hospital that doesn't exist anymore had set two men off on mistaken courses they'd live and die without ever knowing about, and shaped generations to come.

There was the birthdate, the same as Jim's: September 23, 1913. There was the same stamp, skewed at a diagonal: Fordham Hospital. There was the signature of the promising young doctor, Jerome McSweeney, who delivered both babies, assisted by nurses, nurse trainees, and hospital helpers whose identities are now rendered invisible by time. And there was the birth certificate number: 10942. It was one digit off from the number on Jim Collins's birth certificate, 10941.

The little Irish boy, the son of Katie and John Josef Collins, had presumably been born first, and at some point later in the day, the little Lithuanian Jewish boy, the son of Sam and Ida Benson, came along. At a hospital with fewer than a child a day born on average, they were quite possibly the only two babies to arrive at Fordham Hospital that day. And yet somehow, the two newborns had gotten confused.

* * *

In the months after finding Pam and Phylis, Alice searched online and found a book with the ponderous title *Brought to Bed: Childbearing in*

REPORTED DECEASED; D.O.D. 12-11-87

STATE _Ca._ D.C. # 5005

5 H—1913

THE CITY OF NEW YORK,
DEPARTMENT OF HEALTH.
STATE OF NEW YORK.
CERTIFICATE AND RECORD OF BIRTH

No. of Certificate 10942

OF

Name of Child *Philip Benson*

Sex	*Male*	Father's Occupation	*Benson tile maker*
Color	*white*	Mother's Name	*Ben Benson*
Date of Birth	*Sep 28 1910*	Mother's Name before Marriage	*Ida Cott*
Place of Birth, Street and No.	*Fordham Hospital*	Mother's Residence	*468 Benson Av.*
Father's Name	*Ben Benson*	Mother's Birthplace	*Russia*
Father's Residence	*468 Benson Av.*	Mother's Age	*28*
Father's Birthplace	*Russia*	Number of previous Children	
Father's Age	*24*	How many now living (in all)	*2*

I, the undersigned, hereby certify that I attended professionally at the above birth and I am personally cognizant thereof; and that all the facts stated in said certificate and report of birth are true to the best of my knowledge, information and belief.

Signature, *A M Sweeney*

Residence, FORDHAM HOSPITAL

DATE OF REPORT, _____ 19__

THE CITY OF NEW YORK.
DEPARTMENT OF HEALTH.
STATE OF NEW YORK.
CERTIFICATE AND RECORD OF BIRTH

No. of Certificate, 10941

OF

Name of Child *James Collins*

Sex	*Male*	Father's Occupation	*Driver*
Color	*White*	Mother's Name	*Katie Collins*
Date of Birth	*Sept 28 1913*	Mother's Name before Marriage	*Katie Kennedy*
Place of Birth, Street and No.	*Fordham Hospital*	Mother's Residence	*1285 So. Blvd*
Father's Name	*John Collins*	Mother's Birthplace	*Ireland*
Father's Residence	*1285 So. Blvd*	Mother's Age	*32*
Father's Birthplace	*Ireland*	Number of previous Children	*3*
Father's Age	*37*	How many now living (in all)	*3*

I, the undersigned, hereby certify that I attended professionally at the above birth and I am personally cognizant thereof; and that all the facts stated in said certificate and report of birth are true to the best of my knowledge, information and belief.

Signature, *A M Sweeney*

DATE OF REPORT, _____ 19__ Residence, FORDHAM HOSPITAL

The birth certificates for Phillip Benson (misspelled) and James Collins

America, 1750–1950, by the childbirth historian Judith Walzer Leavitt. She bought it and was struck by a picture inside.

The caption read, "Babies on the cart at the Manhattan Maternity and Dispensary, 1912." The black-and-white image showed about fifteen babies piled together, some leaning on each other, like so many cabbages on display at a grocery store.

"Every time I show it, when I give lectures, the whole audience gasps," says Leavitt, a retired professor at the University of Wisconsin–Madison. "You can understand how possible it was to switch babies inadvertently."

Because hospital births were still unusual in 1913, procedures to identify babies were changing and inconsistent. It's hard to know what practices were in place at now-defunct Fordham Hospital, but in 1913, it's entirely possible that Fordham didn't use any kind of identification system at all, that "they just depended on mothers' recognition or nurses' remembrance," Leavitt says. If they did have a system, it may have been crude. I found a number of scholarly papers and patents from the 1920s attempting to address the problem of keeping babies straight in maternity wards: aluminum tags tied around babies' necks, bracelets with lettered beads, slips of paper tied to umbilical cord stumps, and this memorable one: "The child's name written on its chest with indelible pencil." Alas, none of these approaches was foolproof, wrote one prominent obstetrician in a 1928 issue of the *Journal of the American Medical Association*. A "69 could be read 96, or the adhesive plaster labeled J. Jones get on J. James' back, or a necklace intended for baby Flanagan get round the neck of baby van Dorn, if the two babies should be placed together on a table while their beads were being strung," he wrote.

Leavitt told me that it wasn't till the 1930s or '40s that it became standard for hospitals to give both babies and their mothers identifying wristlets or anklets so that they could be matched to each other. Even then, as a few baby-switch cases from the 1940s attest, the methods were far from perfect. Complicating the situation, Leavitt added, was the fact that while some hospitals kept babies sleeping in cots by their mothers' beds, other hospitals kept them in nurseries, increasing the chances of a mix-up.

I couldn't find any pictures of the babies at Fordham from this era, but a 1907 report on obstetrics practice from Bellevue Hospital, with which Fordham was allied and appears to have shared pupil nurses, gives a sense of what giving birth in a hospital could have been like back then. New mothers were wheeled out of the delivery room and into a "lying-in ward," while their babies were weighed and measured and dressed and put into cribs. Newborns were not brought to their mothers to nurse for the first time till they were about six hours old. Could a mother, woozy and exhausted after childbirth—one that may have taken place under the influence of chloroform or ether—reasonably be expected to know that a baby she barely glimpsed six hours earlier was hers?

And let's say she received the right baby. This first period of time, right after childbirth, would not be the last time her baby's identity might be confused. Bellevue's nurses appear to have been handling the babies a lot, given that the hospital enforced a strict feeding schedule of nursing every two hours during the day and twice during the night at precisely 1 A.M. and 5 A.M. And throughout a mother's stay, nurses daily took her baby's rectal temperature, counted his pulse, gave him a bath with white castile soap, dusted him with talcum powder, and washed out his little eyes to prevent infections. Assuming the baths weren't given at the mother's bedside, each one of these daily journeys with the nurse increased the chances of a mix-up.

In their efforts to keep babies healthy, clean, and well-fed, it appears that early maternity wards were unaware of the very real danger they posed. At some point—perhaps while putting the babies into cribs or while parceling them out from a cart—an exhausted hospital employee could not tell a James from a Phillip. Two mothers were handed the wrong babies, perhaps even as they lay near each other in the same large, light-filled ward. Two mothers were handed the wrong babies, and they nursed those babies, and admired their tiny faces, and then took those babies home.

The repercussions of that single event, the act of setting one baby down in one place and another baby down in another place, meant that two men wound up living lives intended for each other. A century later, Jessica Benson would write that this knowledge was humbling. She and her family, and Alice's family, were children of a whim. We are all, of course, children of whims, and we have always known this to be true, long before the DNA age brought to light tales of one-night stands and donor sperm swaps and the mix-up of two babies in a hospital. We are whims of our mother deciding to attend a certain dinner party and meeting our father. We are whims of the sperm cell that beat out millions of other sperm cells to reach our mother's egg. And as we come into being, we are still whims: of childhood accidents, and the precise location of the house we grew up in; of a college admissions officer; of an email never opened; of a drunk driver we never saw coming. As destabilizing as it is to admit, life is just one grand unfolding of accidents—accidents or God's will, depending on how you see it. Yet there is something poignant and deeply disturbing about the idea that an act of momentary carelessness by the professionals charged with caring for the most vulnerable of human beings could have had such profound consequences.

Alice emailed Leavitt and told the historian her tale. Alice doesn't often get emotional, but she does when she asks herself one of the questions she asked of Leavitt then. She wanted to know whether or not Jim ever enjoyed the comfort of his biological mother Ida's arms. "I wanted—at least one time—to have him held by his mother," she told me.

The question has special resonance for her. Alice has three sons, and one day over lunch she told me about the birth of her eldest, Gidian. He's a radiation technologist, and he and his wife and Alice's three grandchildren, whom she adores, live only two houses down. But when Gidian was born back in 1973, Alice told me, she did not know if he would make it. He had to be delivered prematurely by an emergency C-section because Alice was having complications, and afterward, the newborn would sometimes stop breathing. Alice was weak, and the hospital would not tell her what was going on with her son. She remembers finally gaining the strength to walk down the hall to the nursery only to watch in horror as staff crowded around her son to perform some sort of procedure that no one would explain. Bruce was teaching and she could not get ahold of him. The doctor, when he came, could not tell her if her son was going to live or die.

Gidian was three or four days old by the time Alice was finally able to hold him, and this memory of that early separation helps explain why the not knowing and the sadness of that loss—of Ida's loss, of Jim's loss—haunt her.

It is a question without an answer, a puzzle not even DNA can solve.

16
ALICE REDUX

What role did nature play in the lives of Jim and Phillip, and what role nurture? Who were Jim and Phillip really, in their souls, as Pam would put it? And who would they have been if not for that mistake?

"His story is one of a lot of tragedy," Jim's son Jim the younger says of his father. Jim the younger ticks off the losses: his dad's mother dying when he was a baby, his dad going into an orphanage, his dad leaving the orphanage and being on his own before the age of fifteen. What would have been if Jim Collins had lived the alternate life, the one intended for him? The Collins kids can't regret what happened—"If he wasn't switched, we wouldn't be here," Gerry points out—but that doesn't mean that Gerry can't feel sorrow for her father, for what he had at the very beginning of his life and all he was forced to give up.

It's hard to explain this mix of gratitude and regret, Jim the younger says. "Because you loved this person, you wish things would have been a little bit better for them, or a little bit easier," he says. "I'm happy that we know where his family came from, and we met them, and they are good people. It's a little sad, too, that *he* never had that opportunity, and I don't know how he would have felt—years later, after being raised one way, you find out you're somebody different. It's sort of a perplexing situation no matter how you look at it."

Among the photographs Pam showed Alice during The Great Exchange was one of her father, Phillip Benson, as a child. Phillip is about three. His shiny hair is trimmed into a neat bowl haircut, and he is sitting on the back of a pony, lightly gripping the saddle, with a faint smile playing on his lips. Alice scanned and restored the photo and sent it back to Pam, and meanwhile she thought about the image, which provoked in her a pang of jealousy. She did not begrudge Phillip the pleasure of that pony ride, but she thought of her father, given an orange for Christmas, and she wished for him the pleasures of the life he was meant for.

Phillip Benson as a child

It's impossible not to wonder what would have been. Bill Collins wonders if his dad could have had a better education if he'd stayed with his biological parents. "He was a very smart man and he would have been in a different situation," he says. Would Jim have chosen something other than work in regimented institutions that so mirrored life in the orphanage—the army, then correctional work? Might he have done something else entirely, Alice asks, "something where he used his mind?"

Alice and Gerry also wonder what affect the baby swap had on the family that took little Jim into their home. It's difficult to know exactly what was going on during those nine months of turmoil, the nine months between when John Josef and Katie brought their youngest child home from the hospital and when Katie died. Was the turmoil because Katie was sick, or because she and John Josef had separated, or because John Josef was "intemperate"? Were these things related? Most crucially, why did Katie and John Josef separate? For Irish Catholic immigrants to break up in 1913, just after having a child, seemed surprising. Alice and Gerry and Pam all speculate that the separation may have resulted from Katie bringing home a child who did not look like the rest of their family, and John Josef becoming suspicious.

"He had two children already and this was not his child, and he knew it—like a mother seal knows her baby pup," Gerry told me. If Jim had not been swapped, Gerry went on, she believed the Collins family would have remained intact. The theory is impossible to prove, of course, but if it were true, it would mean that the baby swap was not just the engine by which Jim was given to another family, but the means by which that family was dismantled.

It is tempting to wonder what might have been if Jim had known about the switch when he was alive. Would he have been better off? Would it have helped him reconcile with the past, or would it have been a burden for him to contemplate what he'd lost, and threaten the loss of more that was important to him: his memories of his siblings, his sense of himself as an Irish Catholic man with an Irish Catholic family? These *what-ifs* are the existential questions of the DNA age, the ones that loom over revelations of misattributed parentage and donor conception and secret adoptions, over the stories of people like Linda and Laurie and Jason and Rosario and so many others. Who are we, really? Who would we have been if we'd known the truth earlier?

The Collins kids have wrestled with who Jim Collins "really" was and who they, in turn, really are. They have wrestled with whether he would have wanted to know. "He was raised Irish but he was Jewish—that's what I say," Alice says. But her brother, Jim the younger, says, "He was Irish in every sense of the word other than having been born to an Irish couple." And they have wrestled with whether or not they would have told him the truth if they could have. *Would there have been a purpose in telling him?* Alice wondered early on. But eventually, she came around to Gerry's thinking, which is that it would have been the right thing to tell him. "I feel like it would have answered questions for him," Gerry says. "I feel like he would have finally understood." Alice imagines the Collins kids telling their father and Jim saying, as he sometimes did, "Such is life." Still, perhaps it was a mercy that the Collins siblings never had to make that call. "My dad would have lost his identity," Alice says. "He's been kind of spared that."

Are you what you live? Are you what you were born into? As Rosario Castronovo wondered, to what extent can you claim a heritage you didn't grow up with? And how do you decide what to claim? Experience gets a lot of weight, of course, but so do history and intention. Rosario's black grandfather lived and loved as a black man, and paid the price of a black man. The erasure of black identity in his descendants is arguably yet one more loss. Similarly, Jim Collins's biological mother never intended for her son to be raised in another family, in another culture and religion. But if the philosopher Françoise Baylis is right—if identity is something we work out in constant conversation with the rest of the world, a kind of story we tell, shaped by our beliefs and desires and fact-checked and validated by others—then the only person who could have said who Jim Collins truly was was Jim Collins himself.

And likewise, what the Collins kids feel themselves to be varies; it is a question with a seven-sided answer. One brother told me it was a little strange at first to look at himself in the mirror and realize that his own history, and the generations of people leading up to his existence, weren't what he'd believed all his life. John Collins, a practicing Catholic, told me the news had not changed his religious practices, but for Alice and Gerry, I sensed that the revelation felt more personal. Gerry told me a relative had objected when she described herself as half-Jewish, saying "No, you're not, you're Irish," and reminding Gerry that she'd sung an Irish blessing at a wedding. "This doesn't have anything to do with you," the relative told her. But it did, Gerry felt. It had everything to do with her. Gerry told me the news that her father was Jewish made sense, confirming an orientation to the world she had long felt.

What can we make of the fact that some people say this is true for them after DNA revelations, and others don't? Are some people more sensitive to the tiny clues that an out-of-order identity can drop? Or is this a matter of us storytelling human beings looking for patterns to explain the world, the way that Jason saw his own temperament reflected in his father—only to find out later, when DNA set him straight, that this father wasn't his?

I don't know. I do know that Alice says the same thing as her sister.

"One almost wonders whether you have genetic memory for your people," Alice says.

* * *

In January 2017, an unusual group took a cruise to the Caribbean. It was a family reunion, though it was more than that; this assembly of more than thirty vacationers amounted to an illustration of the wonders of consumer genomics, as well as a lesson in how the past lives on in us. It was Alice and all her siblings, plus assorted Collins family members, along with their newfound cousin, Phylis Pullman, and members of her family, and the Collins clan's new swapcuz, Pam. Two years before, many of these people had been strangers to one another; now, they were eating at the buffet, shopping at the ports together, and dressing up in suits and gowns for a huge family photo.

It was oddly comfortable, said Phylis, who didn't want the week to end: "None of them felt like strangers to me because they all looked like they belonged to my family." Someone made T-shirts that said "A Family United: Benson-Collins," and on the back, two trees twisted around each other with a heart in the middle, containing the names "Phillip" and "James" and their shared birthday. One night, Phylis told me, she and some of the family climbed up a whole lot of steps, and the cruise ship turned off its lights, and they stargazed right there on deck, in the middle of the water. "All these stars—you felt you could touch them," Phylis said. The points of light that come to us from years gone by seemed so present at that moment, as if the past had finally arrived.

"It was wonderful," said Pam Benson. "It was like we were all one big swap family." At breakfast one morning, Pam watched Alice's brother John walk into the room and was stunned. There was her grandfather Sam Benson, a man she'd met only in photographs, come to life. "Tears welled up because he looks so much like my grandfather," Pam said.

Even so, Pam told me, when we first spoke about the cruise in 2017, she wished her genetic first cousin Pete Nolan, the man whose DNA had been so crucial to Alice's case, were there. Pam had not met Pete yet at that time, though they'd been emailing. It was hard for her watching Alice find Phylis and those two families come together. She longed for that same experience, what she called the "family bond." And in the years that followed, she told me she also struggled with how to reconcile with a sense that in the swap "my dad got the better end of the deal." That version of events began to bother her. Her father had not asked to be placed in the wrong crib, after all, or on that pony in the park. And his life had not always been easy. He, too, had had difficulties, as well as triumphs.

Phillip Benson had married a woman who was one of four sisters, and the four husbands of the four sisters all went into business together, moving to California to build parts for airplanes. Then Phillip's wife died of cancer, leaving behind two young boys. Phillip remarried and was nearly fifty by the time Pam was born. Pam told me her dad had loved tinkering with cars; he'd buy old Ramblers just so he could fix them up. They lived near Los Angeles International Airport, and Phillip knew the flight schedules in his head, organized by the type of plane. And he kept homing pigeons, a hobby he'd loved since he'd been a child hanging out on a New York City rooftop. Pam remembered driving with her dad to LAX to pick up pigeons he'd bought from all over the country, and she remembered the pretty tumblers, who seemed to fall out of the sky. Her father would put bands on their ankles and drive them miles and miles to free them, and take bets on which birds would come home first.

"He loved them," Pam told me. No matter how far he drove them, the birds always returned. "It was amazing. How did they know to come home?"

Pam told me her father was generous. He had done well in business. A relative moved out to California and Phillip rented the guy an apartment and gave him a job. Every two years Phillip bought a new car, and the old cars went to people who needed them. "Don't worry about it; buy me a beer," he'd say. And then his fortunes turned. He had a stroke when Pam was eleven and stopped working, and the restaurant where Pam's mother had waitressed closed down. Phillip never regained the use of his right hand, and he had to walk with a cane. Phillip lost his oldest son in 1985, and Pam said it seemed that her father never recovered from having to bury his own child. He lived only two years after that, and he was a changed man—quiet, heartbroken.

It had been a few years since the DNA discovery when Pam finally flew out to North Carolina to meet her first cousin Pete Nolan in person. She stepped out of her rental, and Pete was outside taking a stroll, and time stopped while she looked at him. There was so much to see. Her father's nephew had Phillip's height and coloring. And he had those blue eyes she hadn't seen since 1987, when her father had died. Before anyone said "hi," before Pam told Pete to come give her a hug, they were both silent, looking at each other. He was family. She told me she could feel it with her whole being.

They sat at Pete's table and shared stories. Pam pulled out a photograph of her father and told him, "Pete, look at this, you have your uncle's ears." And afterward, Pam kept thinking about what her father's life would have been like if not for the switch. What would his career have been, his loves, his children? "I mean, your mind can just travel and travel," she told me. But there had to be an explanation for it all, right? Pam was raised Catholic, and now she said she could not help but think there was a guiding hand in all of this—in the switch and in its discovery—because, well, because. Maybe because the alternative seemed unthinkable. "This was supposed to happen, because that's what did happen," she said.

Pam told me she'd enjoyed the shock of seeing how very alike Alice and Phylis looked, and she was glad, at last, to have her own version of that story with Pete. She was struck by the fact that two families had been united by a single moment a century before, had taken separate paths, and had wound up back on the same road all these years later.

She told me seeing Pete had given her a sense of closure. She already wanted to see him again. She had trouble putting into words how good it felt to know him after having traveled so far. There was a kind of inevitability to this reunion—as if, like her father's birds, she'd always known where she was going. From the great distance of time, she was finding her way home.

* * *

The trouble with Alice's primary method, using genetic segments to reverse-engineer her family's DNA into the right tree, was not that it wasn't sound. It was that in the years she was using the technique, the genetic testing revolution hadn't yet reached the right people. The databases were too small. In all that time, combing through cousins who appeared to be more related to her than they actually were because of the influence of genetic interrelatedness, and stymied by a preponderance of last names that often didn't follow family relationships in an expected way, Alice never found a cousin close enough to allow her to unravel a relationship.

But in early 2017—about two years after she'd found Jessica Benson and unraveled the mystery of her father—Alice's long years of work with genetic segments were vindicated. A man showed up as a relative at 23andMe. She looked at his chromosomes, and they did not look like those strands of endogamous spaghetti CeCe Moore had talked about. The man shared a whopping 217 centimorgans of genetic material, suggesting a closer relationship than any other Alice had found, and, instead of being arranged into many short, chopped-up segments reflecting relationships far back in time, most of them were large. A big chunk of the shared genetic material was along the X chromosome, suggesting this man was related to Alice through his mother. And that's how Alice found her way to a woman named Ruth Klein—her father's last living first cousin. She flew to New York to meet her.

Looking back on how she solved her case before Ruth came along, Alice is always struck by how it happened. There were so many barriers along the way, and so many accidental connections that almost didn't happen. What if Pete had never decided to go looking for his uncle's family? What if Pete had not written that letter to the orphanage at the precise moment that he did, so that his letter and Alice's letter could cross the desk of someone who read both of them and realized that these two letters were requesting information on the same family? Without that, Pete never would have found Alice, and Alice never would have forged a relationship with the man she still considers her cousin in everything but blood. And without that, she never would have been able to find her biological cousins, because she would not have found Jessica Benson and learned of the existence of Jessica's grandfather, the other baby at Fordham Hospital.

"I keep on thinking, what are the odds?" Alice says. Alice is not really religious, but she can't help but think "there's some overriding force—God? Who knows what?—trying to set this world straight."

I'll add to this pile of improbabilities, although Alice never mentioned these: The whole notion that a person could, by spitting into a vial, uncover the truth of something that happened over a hundred years earlier to two people who are no longer alive—this also seems utterly preposterous, or would have, just ten years ago. Consumer genomics moves so fast that we take for granted now that such a thing *is* possible.

This speed underscores another point, which is that the herculean efforts that Alice had to put forth to find Pam Benson and understand what had happened to both of their fathers would not be necessary now. Because she first tested in 2012, the hunt for understanding took two and a half years. But were Alice testing for the first time just a few years later, she would have been able to journey from the shock and confusion of that unexpected ethnicity estimate to unraveling that her dad had been switched in the hospital and figuring out his identity in just a few days. She would not have needed Pete and Jessica; she would not have needed to test her whole family or send kits out to strangers or have her brother create an app to track 311,467 genetic segments. Finding Ruth Klein's son in the database would have been enough to set the world straight.

17

WHERE WE'RE GOING

In the years between when Alice first spit into a tube and now, the landscape of consumer genomics, and genomics research more broadly, has changed radically. On Twitter, I followed academics, lawyers, and scientists debating many issues touching on genetics research and use. I read books and scholarly articles, knowing that the landscape was changing drastically even in the time it took for those works to be published.

A few months after I flew out to spend time with Alice in Washington State, a huge controversy unfolded over a Chinese scientist who announced he'd created the world's first genetically edited babies, a move that scientists worldwide denounced as unethical, unnecessary, possibly dangerous to the babies, and potentially impacting on future generations. "Ever since scientists created the powerful gene editing technique CRISPR, they have braced apprehensively for the day when it would be used to create a genetically altered human being," the *New York Times* reported. "Many nations banned such work, fearing it could be misused to alter everything from eye color to IQ." One of the inventors of the CRISPR technology, who years before had told a magazine she'd had a nightmare in which Adolf Hitler himself asked her for a tutorial on "this amazing technology," pronounced herself "physically sick" when she learned the news about what had happened in China.

As I was reporting and writing, I followed many growing debates: about so-called three-parent babies, created by a technique called mitochondrial transfer to avoid the passing down of certain genetic disorders; about the potential and ethics of growing efforts to conduct genetic sequencing of newborns in hopes of spotting otherwise undetectable health issues; about research attempting to link genetics to IQ, to educational attainment, and to income; about work being done within the field of epigenetics. On that last topic, there are even debates over how the term "epigenetics" itself is used; the epigenome, according to the National Human Genome Research Institute, "is a multitude of chemical compounds that tell the genome what to do"; it can be affected by the environment, and in some cases, epigenetic changes can be inherited. But how much can the epigenetic influence of trauma, for instance—experiences like war and poverty—be passed down? Some studies have suggested "that we inherit some trace of our parents' and even grandparents' experience, particularly their suffering," the *New York Times* reported in late 2018, but critics "contend that the biology implied by such studies simply is not plausible."

And meanwhile, as I was talking to seekers, the prospect of whole genome sequencing was becoming ever more possible for everyday Americans. The Human Genome Project took thirteen years and cost $2.7 billion, but in recent years the cost of sequencing has dropped to about one thousand dollars or less, and by the time I was writing this book, Illumina was talking about a not-too-far-off future in which it will cost one hundred dollars, about the same as the current cost of consumer tests like AncestryDNA. In *She Has Her Mother's Laugh*, science journalist Carl Zimmer writes about having his own genome sequenced a few years ago. "When I started writing about science, if someone had said, 'Do you want to get your genome sequenced?' it would have been like saying, 'Do you want to go to another galaxy?'" Zimmer said in a talk. He took his raw data around to different scientists for analysis, and was struck both by what this data could tell him and by what it could not. Studying the actual Neanderthal genes he'd inherited was amazing, Zimmer said: "I can say that I got that gene from somebody sixty thousand years ago, maybe, and that does give you this profound connection to the past." But at the same time, for so many genes he was looking at, "we don't really know what most of them actually do."

"What if: Everyone Had Their Genome Sequenced at Birth?" was the title of a panel discussion at the World Economic Forum Annual Meeting at Davos in early 2019, conjuring up images of a future that several geneticists and marketplace experts had been describing to me in which genomic data is as integral and accessible to doctors as a person's health history and BMI. The audience was split on whether such a thing should happen now, but nearly unanimous that it was inevitable that most infants would ultimately be sequenced. The panelists debated the benefits and drawbacks. In cases of children known to be ill, a strong case could be made for sequencing, but were we ready to start uniformly sequencing healthy babies, and what to do when genetic issues cropped up for which nothing could be done? What were the implications for consent, privacy, economics, and health inequities?

I watched as the use of genetic genealogy to narrow in on criminal suspects went from a controversy to a novelty to a fairly normalized aspect of crime-fighting, even as bioethicists, lawyers, and legislators were still debating it. A woman named Brandy Jennings, who lives in Alice's town of Vancouver, Washington, found out the genetic file she'd uploaded to GEDmatch and forgotten about had been used to locate a murder suspect, a second cousin once removed whom she'd never met. She said she was glad her DNA had been able to help, even if she never could have anticipated it.

And I followed genetic genealogy blogs explaining the new techniques unveiled for "smashing through brick walls" by companies like Ancestry and MyHeritage at RootsTech 2019, the massive Salt Lake City conference put on annually by FamilySearch, which boasted twenty thousand attendees that year. The features do things like show which common ancestor likely connects two DNA cousins, and sort DNA matches into clusters. In short, they make easier the kinds of breakthroughs Alice could only dream of when she first started trying to unravel the mysteries of her DNA.

The pace of change was so rapid that I became something of a Twitter junkie, nervously checking many times a day to make sure I wasn't missing anything. I worried sometimes that writing this book was a fool's errand—that whatever happened in the realm of genetic privacy, forensics, consumer protections, and family relationships that seemed fresh and revelatory as late as 2019 would seem familiar and even obvious to readers in 2020.

But then I would remember that I was in a bubble. The people who were getting DNA kits for Christmas had no idea what was coming for them. And the ramifications of what they might find would not be short-lived; rather, they amounted to a fundamental reshaping of the American family. It was something they would deal with for the rest of their lives and pass on to the generations that followed.

* * *

Every morning, Alice has her routines. She wakes early and puts on MSNBC, plays solitaire, does several crosswords on her iPad, and checks the DNA databases, to see if any new DNA cousins have shown up. She makes notes of how much genetic material she shares with each one, and how they might be related, and reaches out to the closest new cousins. Her curiosity and her desire for connections means she's constantly emailing and calling and driving and flying to visit with an ever-expanding set of relatives. Because of her love of genealogy, Alice's family is bigger, and her world smaller, than many people's. Among the things her father passed on to her is a sense that family is everything.

Alice being Alice, she is still searching and has built a huge family tree she calls Twisted Branches, which incorporates both the Benson and Collins lines, both the family her father knew as his, and the genetic family he never knew. The tree has 1,331 people in it, with the earliest ancestor born in 1536. Alice has made contact with a relative in Israel, and found cousins in South Africa, and discovered links to a family of Estonian rabbis. She used her genealogical knowledge and her trick of thinking sideways to figure out her likely connection to a DNA cousin through a tiny village in Belarus. When George H. W. Bush died, Alice sent me a photo taken of his casket—the forty-first president was a seventh or eighth cousin of her mother's, something she uncovered by tracing back to the Plymouth Colony—flanked by a marine who happened to be her dad's great-grandnephew. "We never thought we'd see a state funeral where both sides of our family were represented!" she wrote. The genealogy she was doing, evoking long-gone towns in eastern Europe, and long-dead family members with names like Freyda and Yetta and Mendel, meant that the past was not a static thing, any more than Jim's identity was. It was a puzzle forever being solved. In asking all those questions, in combing through old archives, Alice was continually putting her father's story together, keeping him with her.

Alice fields questions from people trapped in genetic mysteries, and skipped a class in the finer points of serger sewing to tell her story at a nearby genealogical society. She has helped Pam unravel her genealogy, and together the swapcousins have been trying to get New York State to annotate their fathers' birth certificates, to reflect their true parentage. She continues with her sewing projects and made a holiday runner for Hanukkah, which she has added to the long list of holidays her family celebrates. She told me that sometimes she goes to a reform synagogue for Friday-night services. She never had been a "good Catholic," by her own reckoning, and finding out her father was Jewish by birth has given her another means to feed her spirituality.

One afternoon, Alice showed me something amazing. We were sitting in her office eating peanuts and M&M's. She went onto GEDmatch and into a tool called Lazarus. She showed me how, using the DNA tests she, her siblings, and father's cousins had taken, she was able to get a sense of what her father's genome would have looked like. She didn't need his saliva; Jim had seven living children, and each had inherited genetic material from him, some of it the same and some different. Alice verified which of the Collins siblings' genetic segments came from their father by matching them against known paternal cousins, and, by putting it all together, she could approximate a good portion of what Jim's chromosomes looked like, effectively raising him from the dead.

"God*damn*," I said, wondering for the umpteenth time if I'd ever met anyone so nerdy and so cool.

* * *

The fundamental lesson of the DNA age is that the past is not over. We may feel we are leading modern lives, having left behind certain tragedies and injustices, certain mistakes and anachronisms, but it is all still there. It is etched into us, and it only requires technology to be revealed. It is revealed in the segments of DNA that demonstrate how contemporary African Americans carry the legacy of white men who forced themselves on the women they enslaved. It is revealed in the discoveries of those who were donor-conceived at a time when such reproductive methods were kept secret, even within families. It is revealed in cases of adoptions and NPEs, some covered up when doing so was a matter of survival. DNA reveals decisions made in the aftermath of events of all kinds: acts of passion, acts of violence, acts of love. We get to decide who we are, but DNA allows us to inform that decision by understanding the truth of how we got here, of how we came to be. It forces us to look back and examine the circumstances under which difficult and life-altering decisions were made. It gives us context.

When I interviewed Bennett Greenspan of FamilyTreeDNA in Texas, he told me he could not have envisioned what his business would become. During that weekend in August 1999 when he had the idea of using DNA to get around a brick wall in his own research, he was not imagining how DNA could be used to explain how we all fit into the massive family tree of our species, or anticipating the FBI uncovering the identity of suspects based on the genetic information of fellow hobbyists in his database. He was not thinking that one day genetic testing could be used to show that two babies had been switched in the hospital back in 1913. He was just a seeker. He was "looking for my *emes*," he said, using a Hebrew word. "My real truth."

But of course, your tree is my tree is his tree is her tree. Your emes is not my emes, and sometimes, in a genetic database, they collide.

Things have changed so much since the advent of consumer genomics. Alice's search encapsulated a brief moment in time when the technology existed, but the databases were relatively small, meaning that the techniques of genetic genealogy made unraveling connections just barely possible, if combined with enough work and luck. But in the future, we may not even remember this moment in time, even as it is truly the moment when everything changed, and when a woman like Alice could make the choice whether or not to search. Because in the future, and I'd guess it's a near one, enough people will have undergone consumer testing that solving a genetic mystery will be in many cases almost instantaneous. And nearly everyone related to the seeker—which will be not a special class of people willing to learn complex genetic genealogy techniques, but, effectively, anyone who sends in their saliva—will be rapidly implicated in any discoveries that arise, whether or not they choose to test. When the future catches up to the past, and there are no genetic secrets, we will need to rethink many things—starting with how we talk to our children.

In the summer of 2018, CeCe Moore told me that since Alice's discovery, she'd assisted in solving five more hospital baby-switch cases. Alice's father's switch had occurred the furthest back in time, and the most recent had taken place in the 1950s. News stories have recounted two instances of baby swaps uncovered through recreational testing, both involving girls born in the United States in the 1940s. Both were solved quite rapidly because the databases pulled up a biological sister in one case, a mother in the other. There was no quest, as there had been for Alice—just a sudden, searing revelation.

"Sometimes I had that sense that I didn't quite fit in with them," said one of the women, who was switched at a Minneapolis hospital in 1945, reflecting back on the family she was raised in. She was echoing themes I'd heard some seekers give voice to, a lifelong sense that not all the puzzle pieces of their nuclear families fit together, even when they didn't know why. But women in both stories also talked of an expanding of their sense of family once the revelation was made, rather than a replacement of it—not nature over nurture, not a zero-sum game. It was language I'd heard from other seekers, like Laurie Pratt, who never told her father that she didn't share DNA with him because her dad was her dad, no matter what a test said.

Throughout the years I was researching and writing this book, my interest in DNA was not just professional. My parents both began researching their roots in earnest, helped along by growing DNA databases and genealogical archives. My mom was able to confirm the identity of her great-grandfather. My dad met the son of a first cousin he never knew along one branch of his tree, and along another, he found a genealogist relative in Sweden, and we decided to embark on our own DNA heritage trip. In southern Sweden, we visited the farm where my dad's grandmother Sigrid grew up. We walked the lanes Sigrid walked and saw the church where she worshipped and the school she attended, learning about the entrenched poverty and family dynamics that must have contributed to her decision to emigrate at sixteen. We shared meals with my dad's second cousin and her daughter, people we never would have known but for ancestry testing. We became, as a family, seekers, the way so many Americans are becoming. I fielded messages from a woman trying to help an adoptee who wondered if my mother's DNA might hold the key to the adoptee's identity, and I scrutinized the overlapping segments with this distant cousin, and thought endogamy might be confusing the issue. I scrolled through my match lists, pausing at the close cousins with unfamiliar names, wondering when the roulette wheel of some unexpected revelation *would* stop at our family, because by now it seems inevitable that such a thing will happen to someone in our familial orbit, eventually.

I spoke again with Rosario Castronovo, who described his search for the truth as an effort to reconcile with decades of his life during which the most important things were left unsaid. I spoke again with Linda Minten, who told me she's been helping other seekers find their genetic families, and she warns them to hope for the best but be prepared for the worst. "Some people bite and some are open-armed." I emailed with Ricki Lewis, the geneticist and science writer; less than two months after our first conversation, she'd discovered she had not just one half-sibling through the sperm donor she'd never known was her genetic father, but six. AncestryDNA kept spitting out new ones. She was a "DC-NPE" in the lingo of this world—the product of both donor conception (DC) and a "not parent expected" event. Lewis is the author of a college textbook on human genetics, and she recently revised the thirteenth edition to include a chapter on NPEs, informed by her own experience. "Bioethics seems to be lagging behind the science, dumping a mess onto the mental health community," she wrote in a recent column on ancestry testing.

And meanwhile, the headlines kept coming—"New Siblings, Old Secrets," and "Your Father's Not Your Father," and "My DNA Test Revealed a Genetic Bombshell," and "Woman's DNA Test Revealed a Shocking Family Secret"—and countless tales about surprises for "adopted Dover woman" and "Tuscaloosa dentist" and "Houston-area mother" and "Minnesota man." And the seekers kept flooding the Facebook help sites, and the children of NPEs kept searching social media for clues about their biological fathers, and the adoptees kept crafting careful letters of introduction, and the genealogy addicts kept building out their trees, and a whole swarming mass of Americans found themselves converts to a search many never planned to embark on, a question they'd never known they were asking:

The question of themselves.

SELECTED BIBLIOGRAPHY

I relied on many sources to understand the broader scientific and historical themes I was interested in exploring in *The Lost Family*, most notably five excellent books: Carl Zimmer's *She Has Her Mother's Laugh*, David Reich's *Who We Are and How We Got Here*, Christine Kenneally's *The Invisible History of the Human Race*, Siddhartha Mukherjee's *The Gene*, and Adam Rutherford's *A Brief History of Everyone Who Ever Lived*.

François Weil's *Family Trees* offered the history of American genealogy I relied on for the book, and *Hearts Turned to the Fathers* recounted the great influence of the Church of Jesus Christ of Latter-Day Saints on the field.

For the early history of genetics and the rise of eugenics recounted in chapter two, the following books by historians of science and other scholars were invaluable: Peter J. Bowler's *The Mendelian Revolution*, Nathaniel Comfort's *The Science of Human Perfection*, Nicholas Wright Gillham's *A Life of Sir Francis Galton*, Daniel J. Kevles's *In the Name of Eugenics*, Mark A. Largent's *Breeding Contempt*, and two books by Paul A. Lombardo: *Three Generations, No Imbeciles* and *A Century of Eugenics in America*. I also relied on research and work by Gregory Radick, Susan Rensing, and Alexandra Minna Stern.

To understand the cultural meaning of DNA, as well as our cognitive biases about it, and how popular portrayals of it intersect with our thinking about race, I drew on several books, including *The DNA*

Mystique, by Dorothy Nelkin and M. Susan Lindee; Steven J. Heine's *DNA Is Not Destiny*; and Alondra Nelson's *The Social Life of DNA*.

Thomas H. Murray's *The Worth of a Child* offered a fascinating exploration of timeless moral and philosophical questions about the relationships between parents and children.

Genetic Secrets, edited by Mark A. Rothstein, explores themes around genetic privacy we've been grappling with for decades.

Erin E. Murphy's *Inside the Cell* delves into privacy, accuracy, and justice in the use of forensic DNA.

For a history of early maternity wards, I relied on Judith Walzer Leavitt's history of childbearing from 1750 to 1950, *Brought to Bed*. Sandra Opdycke's book *No One Was Turned Away* helped me understand New York's public hospitals, and Evelyn Gonzalez's *The Bronx* fleshed out the social and historical context for immigrants living near Fordham Hospital after the turn of the twentieth century.

Dani Shapiro's memoir *Inheritance* movingly recounts the experience of a genetic surprise from the inside out. Other books that shaped my understanding of the power of DNA testing and genetic genealogy both to disrupt and to reunite are *Finding Family*, by Richard Hill; *The Stranger in My Genes*, by Bill Griffith; and *The Foundling*, by Paul Joseph Fronczak and Alex Tresniowski.

In addition to other research and reporting, I consulted Ricki Lewis's comprehensive and comprehensible college textbook *Human Genetics*, and for pure pleasure reading on the topic, I recommend *The Cartoon Guide to Genetics* by Larry Golnick and Mark Wheelis.

Those hoping to understand how to use genetic genealogy to find their biological families—as well as how to reach out to those families, and what to expect—should seek out the very fine guide *The DNA Guide for Adoptees*, by Brianne Kirkpatrick and Shannon Combs-Bennett. Blaine T. Bettinger's *The Family Tree Guide to DNA Testing and Genetic Genealogy* offers a broad and exhaustive overview of the field.

ACKNOWLEDGMENTS

I am deeply grateful to Alice Collins Plebuch, her sister, Gerardine Collins Wiggins, and their brothers Jim, John, Bill, Brian, and Ed for allowing me into their lives and helping me understand their emotional trajectories, as well as the experiences of their father, Jim Collins. Alice's incredible mind and perseverance made it possible for her to solve the mystery of her father; her record-keeping and meticulous memory—not to mention her time and patience—made it possible for me to recount her efforts. Gerry offered important context and emotional perspective, and had records and recollections that deepened my understanding of the family and its search. I am also deeply grateful to Pam Benson for her time and thoughtful insights into her own journey and the life of her father, Phillip Benson.

I want to thank the "seekers," many of whom spent hours telling me their stories of heartbreak, wonder, and love. It is a leap of faith to tell a stranger your shock, your shame, your anger, and your fears, a leap of faith to let a stranger into your most intimate musings. I found myself grateful for their trust and determined to tell these stories with as much sensitivity and emotional nuance as possible, in hopes of doing justice to a phenomenon that is increasingly a part of our American story. One of the seekers I interviewed said she hoped that accounts like hers will make genetic surprises, and the secrets they reveal, "not so taboo." That's my hope, too.

This book began when my editor at the *Washington Post*, the wonderful Amy Argetsinger, came to me with the idea of writing a feature story about DNA revelations. She and others at the *Post* gave the story the space, design, and platform to catch the attention of readers—a lot of readers. As hundreds of emails poured into my inbox with captivating stories that made me gasp, smile, and cry, I knew I wanted to write a book that would attempt to capture the scope of how this technology is changing how we relate to one another. *The Lost Family*'s other fairy godmother is the wise and thoughtful genealogist Jennifer Mendelsohn, who first suggested to Amy that she commission a story on this important topic. Jennifer's guidance about and enthusiasm for the world of genetic genealogy have been invaluable to me throughout the process of writing this book, as was her reading of my manuscript early on. I am grateful for her foresight, patience, and feedback.

A number of experts have generously given their time to help make this book as accurate and compassionate as possible. Genetic genealogist CeCe Moore first told me Alice's remarkable story in early 2017. Over many subsequent conversations, CeCe shared countless insights, much technical knowledge about her field and her work as a search angel, and a big-picture, sociologist's-eye-view of the changing state of the American family. I am exceedingly grateful to geneticist Ricki Lewis for reading through the manuscript to make sure the science was correct, comprehensive, and comprehensible. Her book *The Forever Fix: Gene Therapy and the Boy Who Saved It* is a model of compelling science journalism. Genetic counselor Brianne Kirkpatrick counseled me on the science and ethics of her field, fielded countless questions from me about everything from *BRCA1* mutations to high runs of homozygosity, and spot-read certain science-heavy paragraphs. Genetic genealogist Laurie Pratt McBriarty-Sisk acted as a sounding board and sensitivity counselor, helping me understand experiences like hers from the inside out, and spent an inordinate amount of time making sure I understood the techniques involved in an unknown parent search. Genetic genealogist Debbie Kennett has always found time to answer my questions about her field and the home DNA testing industry, and she's a font of knowledge about ethics, privacy, and international issues.

I also want to thank Kathryn Garber, Emory University professor of human genetics, who read and critiqued several DNA explainer paragraphs that appear early in the book, as well as the American Society of Human Genetics for facilitating this assistance. I am indebted to Johns Hopkins science historian Nathaniel Comfort, whose guidance about the early history of genetics (recounted in chapter two) was invaluable, as were his recommendations about the best historians and sources to consult. I'm grateful to University of Leeds historian of science Gregory Radick, who coached me through the nuances of technical language and aided in my understanding as I worked on this same section. Many more people contributed to *The Lost Family*; in cases in which I've quoted someone at length, you can usually assume that meant one or more long initial interviews, numerous follow-up emails, and another epic fact-checking phone call at the tail end. I also want to mention those who don't appear in the book much, if at all, yet lent a special expertise or were especially supportive of the book and its mission. These include Megan Allyse, Alexander Beider, Benjamin Berkman, Shai Carmi, Ellen Wright Clayton, Ken Cobb, Michael Coble, Robert Cook-Deegan, Graham Coop, Michael "Doc" Edge, Colleen Fitzpatrick, Dov Fox, Tana French, Ariella Gladstein, Evelyn Gonzalez, Jason Harrison, James Hazel, Erika Hutchcraft, Razib Khan, Maarten Larmuseau, Thomas Murray, Sandra Opdycke, Itsik Pe'er, Andelka Phillips, Julia Robbins, Judy Russell, Alicia Semaka, Dani Shapiro, Alexandra Minna Stern, and Carl Zimmer.

My agent, Jess Regel of Foundry Literary + Media, believed in this book when it was just an idea and has offered me savvy counsel and support at every step. I have been thrilled and honored to have as my editor Jamison Stoltz, whose unerring sense for narrative and pacing helped me see how I needed to shape the manuscript. Like *The Lost Family*'s protagonist, Alice, Jamison can see the big picture at the same time as he can tell you where the "nits" are. I am grateful to the whole team at Abrams Press for their support and guidance, including Lisa Silverman, Gabby Fisher, Kimberly Lew, and Sarah Robbins.

Many friends have been with me through this marathon, cheering me toward the finish. In particular, I want to thank Ilene Wong, also known as the young adult author I. W. Gregorio, who since becoming my friend in college has encouraged my ambitions and pushed me to take up more space. Ilene's *why-not?* attitude is infectious. Karen Pasternack knows me so well that at times I'm certain we share a brain. She has comforted and counseled me countless times during the course of this book, making me feel braver, calmer, and wiser. And Nina Morgenlander's faith, love, and optimism have always made me feel like I can do anything. Many other friends, family members, colleagues and wise people offered encouragement and advice, including Angela Giuffra, Monica Hesse, Mark Kuniholm, Katherine Reynolds Lewis, Henry Miller, Suz Redfearn, Gina Scharoun, Lyn Traverse, Neely Tucker, and Gustavo Stolovitzky.

Writing this book has caused me to think a great deal about inheritance, genetic and otherwise. I am deeply grateful to my mother, Sarita Eisenstark, whose curiosity, encyclopedic knowledge and intellectual rigor made me a journalist before I was one. And I want to thank my father, Charles Copeland, whose emotional honesty, generosity and optimism have helped shape my compassion and moral compass, invaluable skills for a writer.

To my children, Olive and Lev, thank you both for your wisdom, sensitivity, and love. As a family we worked together to make this book real. And to my husband, Dan, thank you for believing in me before I did; thank you for your calm and certitude and soulfulness; thank you for always making time for this project, even as your own load grew heavier. I wouldn't want to attempt to write a book (or do anything else, for that matter) without you.

INDEX